D0173235

Mind into Matter

A New Alchemy of Science and Spirit

Fred Alan Wolf, Ph.D.

Moment Point Press

Mo. ent Point Press. Inc.
P.O. Box 920287
Needham, MA 02492
www.momentpoint.com

Cover photo © Krister Nyman
Cover design by Metaglyph
Typeset in AGaramond
MG

Illustration on page 19: "Peering through the cosmic sphere"

by Camille Flamman, 1888.

ISBN 0-9661327-6-9

Printed in the United States of America

Also by Fred Alan Wolf

Matter into Feeling
The Spiritual Universe
The Dreaming Universe
The Eagle's Quest
Parallel Universes
The Body Quantum
Star Wave
Taking the Quantum Leap
Space-Time and Beyond

Contents

Introduction	Awakening the Mystery	1
Chapter 1	Void: The Impossible Life/Death Principle	13
Chapter 2	The Word: Something from Nothing	25
Chapter 3	The Mind in Body: The Desire to Move	37
Chapter 4	Inertia: The Mysterious Resistance	51
Chapter 5	Life: The Body in Mind	65
Chapter 6	Endless Fertility: Is the Force with Us?	85
Chapter 7	My Time Is Your Time: Anything is Possible	101
Chapter 8	Meaning and Manifestation: The Great Gathering	117
Chapter 9	Structure and Beauty: Spirit and Soul	129
Appendix	Existence: Mind in a Box	141
Endnotes		153
Bibliography		165
Index		171

Introduction

Awakening the Mystery

As a scientist and writer I am often concerned with how to offer new, speculative, and exciting concepts to my readers while staying true to my profession as a scientist. Apparently scientists who write books are expected to stay within certain limits of respectability; they should not stray too far from what their peers accept as established dogma. If they do, they are likely to be dismissed as cranks or just plain kooky.

But today as we enter a new millennium we are also entering a whole new way of existing in the world. The modern computer, the advent of quantum computers, breakthroughs in biology, high-speed global traveling, and near-instantaneous communication have opened up wide ranges of human knowledge. People from the various scientific, religious, and philosophical disciplines have begun building bridges between science, spirituality, shamanism, ancient magical practices, metaphysics, and the functioning of the human body, among other areas. So many bridges are being constructed, in fact, that it is difficult to determine just what we should believe. Should we only read and accept what card-carrying scientists tell us? Perhaps we should accept only the words of Nobel, Pulitzer, and other prestigious prize winners. Good sense tells us that if we do, however, we are in deep trouble, for often these writers are no better than the average person when it comes to imaginative or speculative venturing. Worse yet, sometimes even the best minds become far too conservative or far too prejudiced.

While I am not saying that we should dismiss so-called "great minds'" attempts to explain their ideas to a public eager to have them, I do say that there is much room for good, inspired speculation by scientist-writers such as myself, who in explaining the workings of science also offer their unabashed vision of what's to come—even if that vision takes us far beyond the borders of acceptance, and particularly if this vision offers some basis for hope and inspiration.

In this speculative and imaginative book I attempt to go further than I have gone before by offering new ideas based on some ancient visions. The old alchemists, in their attempts to make sense of the world, alter it, and discover its magical secrets, first brought forward the seeds of these ideas. Today, the modern form of these same ideas arises from quantum physics, neurobiology, and information theory. Such concepts deal with human beings, their minds and bodies, and their attempts to control, alter, and cope with their environments, whether those environments extend as far out as a distant galaxy or are as close as their own hearts and brains. The goal of modern scientists echoes that of the ancient alchemists.

Ancient Alchemy

Old legends preserved by authoritative teachers of Judaism assert that the angel at the gate of Eden instructed Adam in the mysteries of both Qabala and alchemy. In fact, the tenets of alchemy, Hermeticism, Rosicrucianism, and Freemasonry are all inextricably interwoven with the theories of Qabalism.[1] And, they all had one common goal: the transformation of the base or common into the pure or rare. Or, to put it simply, *transforming mind into matter.*

Qabalism greatly influenced medieval thought, both Christian and Jewish. It taught that within the sacred writings there existed a hidden doctrine, which was the key to those writings. Eventually, however, the simple Qabalism of the first centuries of the Christian Era evolved into an elaborate theological system, which became so involved that it was next to impossible to comprehend its dogma.[2] Possibly, alchemy and Qabalism split off here. Certainly we can date the principles of alchemy back in time to the ancient Egyptians, for whom it was the master science. The Chaldeans, Phoenicians, and Babylonians were also familiar with the principles of alchemy, as were many people of the Orient. It was practiced in ancient Greece and Rome, and during the Middle Ages it was a science and a religion as well as a philosophy. Often seen as rebels against the religion of their day, alchemists would hide their philosophical teachings under the allegory of goldmaking. In this manner they were able to continue their art and ways, receiving only ridicule rather than persecution and death.

Most modern dictionaries popularly dismiss alchemy as an immature, empirical, and speculative precursor of chemistry, having had as its object the transmutation of base metals into gold. But, although chemistry did evolve from alchemy, the two schools of thought never really had much in common. Whereas chemistry deals with scientifically verifiable and objective phenomena, the mysterious doctrine of alchemy pertains to a hidden, subjective, abstract, and higher order of reality. This reality constitutes the basis of all truths and all spirituality. Perceiving and realizing this reality is and was the goal of all alchemists. They called this goal the *Magnum Opus* or Great Work—the Absolute Realization. It was seen as the Beauty of all Beauty, the Love of all Love, and the Highest High. To witness it required that consciousness be radically altered and transmuted from the ordinary (lead-like) level of everyday perception to a subtle (gold-like) level of higher perception, so that every object is perceived in its perfect archetypal form—the Absolute, the Holy of all Holies.

This transmutative process, the *Magnum Opus*, is at one and the same time, both a material and a spiritual realization. This fact is very often overlooked. Some commentators claim alchemy to be wholly a spiritual discipline, while others seem interested only in finding out whether gold was actually made and by whom. Both attitudes are misleading. It is essential to keep in mind that there are precise correspondences, fundamental to alchemical thought, between the visible and the invisible, above and below, matter and spirit, planets and metals.

In his book, *Transcendental Magic*, Eliphas Levi wrote:

> The Great Work is, above all things, the creation of man by himself, that is to say, the full and entire conquest of his faculties and his future; it is especially the perfect emancipation of his will, assuring . . . full power over the Universal Magical Agent. This Agent, disguised by the ancient philosophers under the name of the First Matter, determines the forms of modifiable substance, and we can really arrive by means of it at metallic transmutation and the Universal Medicine. [3]

The processes of "the creation of man by himself" begin with a primary or archetypal image of that *man*. Creating this image requires some doing. It appears to me that we must use symbolic tools to do so. I have discovered that the Hebrew letters themselves are just the tools needed. Cris Monnastre, in her introduction to the fifth edition of Israel Regardie's *The Golden Dawn*, explains:

> . . . I would suggest . . . begin the task of memorizing the Hebrew alphabet. Within this system, the Hebrew alphabet has no connotation of religion or sect. Its letters are consid-

> ered "generic" and "holy" symbols—powerful doorways into the inner world—and are not associated with dogma or esoteric religious organization.[4]

The ancient mystics first saw these Hebrew letters, these symbolic doorways, as a universal code and thus they set out to completely grasp their meaning. Their goal was to create the image of the primordial human being, and to do this they had to allow the symbols to come alive within them and connect, providing new insights into spiritual and material existence. If they succeeded, they would become fully realized human beings.

This realization comes directly from the Biblical word that man and woman were created in the image of God. Hence, divine life must exist in the human being; and this divine existence must appear and be realized in each part of the human body. A fully realized mystic then becomes, in the image of God, the *Adam Kadmon* (אדם קדמון). According to the mystics, from this Adam all human life originates.

Throughout its long history, Qabalism has attempted to tie two worlds or stages of human development together. The first world is primitive mythology and the second is spiritual revelation.[5] To attempt becoming spiritually enlightened without realizing the world of mythology within us is a serious mistake. People who attempt this often find themselves "in battle with the devil" or "in fear of evil." Carl Jung referred to this mythological avoidance as the "shadow." Isaac Luria's sixteenth-century school of Qabala based in Safed (in what is now Israel), clearly emphasized this. According to Isaac Luria, creation began when God withdrew Himself into Himself in an impossible to imagine self-referential loop.[6] From this withdrawal a divine light emanated and flowed into the first space ever to exist. Our own three-dimensional space was a later development of this primordial space. And the Adam Kadmon—the first being—came from this light. From his eyes, mouth, nostrils, and ears, unconfined primal light emanated. In a great overwhelming mystery, special vessels containing this primal light then appeared out of nothing. These vessels were primal or seed-like matter. But the primal material vessels broke, and chaos was liberated. From this, ultimately, man fell into space-time as a kind of mental projection of the Adam Kadmon.

Creating a New Vision out of Science and Spirit

And so today the mysteries still persist. As smart as we are in the modern world, we apparently can never pass behind the veil which divides the seen from the unseen except by engaging ourselves in the way appointed by the ancients—*the Mysteries.* The questions are as vivid today as they were to the early minds that first thought them. What are we? What is intelligence? What is our source? What is the point of Life? We still look for the tools of our personal transformation. Self-help books fill our shelves. And even with our material needs covered, many of us feel lost and hopeless, driving our way through an objectively stuffed universe with a vacancy in our hearts.

Did the ancients answer these questions? Who are we to say that they didn't? With our modern, "objective" science-oriented minds, are we even capable of understanding the discoveries of, the wisdom of, the ancient alchemists—even if it's right before us? Physicist Wolfgang Pauli once put it that scientists went too far in the seventeenth century when they attempted to make everything understandable strictly as objective science. By denuding the subjective view from any firm ground, much was lost. In much the same way that modern dictionaries make alchemy a mere shadow of the chemistry to come, modern science has attempted to make the study of the subjective a mere reflection of the objective and reducible science of matter. Some of us, including many scientists, don't agree with the new objective materialism. We believe in our heart of hearts, as did the alchemists that came before us, that something far richer than materialism is responsible for the universe.

So, can we in the modern world pass beyond the veil? In this book I affirm that we can. That armed with the ancient knowledge and the modern vision that comes from modern physics, particularly quantum physics, we can rediscover what the ancients may have known. All we need are a few basic concepts—a new way of seeing the old way. I have given a name to these new ways of seeing; I call them the *new alchemy*. So, perhaps we can call ourselves *new alchemists.*

You can certainly think of me as a new alchemist. Indeed, I find myself in complete sympathy with my ancient forebears. As I search through my memories, many recollections of this interest flood my mind. I realize that I have always been interested in magic and transformation.

I remember a particular day when I was playing in the front hallway of my apartment building. I was barely eight years old. I stood at the top of the stairwell and looked down wondering if I could fly down the nineteen or twenty stairs reaching to the ground floor from our first floor apartment. Without thinking, I skidded down the stairwell with my feet only barely touching the leading edges of each step. I was on the ground floor in a flash, and I had not slid down the banister, nor had I placed my feet on any of the steps.

When I grew older and remembered what I had done that day, I realized it was impossible. My feet just were not long enough to go from one step edge to the next without my falling flat on my face. Was this just a dream of super powers, or had I actually skidded down those stairs?

Throughout my early years I maintained my interest in magic and fantasy. That interest carried me into thinking about the world a little differently from my fellows. It led me into quantum physics and to my eventual writing of this book.

I am certainly not alone. I want the reader to realize that today, just as thousands of years ago, many individuals are attempting—sometimes together and sometimes alone—to discover the magical, arcane solution to the enigma of the universe. They seek a hidden, abstract and higher order of reality that would include the subjective as clearly as it does the objective.

This Book and the Story of You

In *Mind into Matter* we will explore how the mind enters into the body at the cellular, molecular, and neural-molecular levels and becomes ensnared, almost—though not quite—believing that it is the body. The reality that mind dimly senses itself as beyond the body will offer a new insight into how the mind and body work as elements in an alchemical laboratory. Like no lab you may have ever realized, the alchemical lab appears very naturally in the world of our dreams and preconscious thoughts. In this lab we learn to develop a magical but ever-movable boundary, called real/imaginal, that divides our mind-body into separate selves, selves which appear to be body images in a real world around them.

Then, we will learn how to conduct experiments at the frontier of the real/imaginal realm. The outcomes of these experiments will result in new information and new transformational possibilities. We will experience this information entering our dreams and, possibly more importantly, during our waking thoughts. And this will lead us into a new vision of life and time. We will see how our brains act as time machines reaching into the future to obtain information and into the past to confirm the validity of these data. We will see how meaning arises from this future-to-present-to-past-to-present information transfer; and how this meaning alters and changes what we believe and what we experience manifesting physically in the world, both personally and globally. And finally, we will complete our journey with a new vision of mind, body, spirit, and soul, and a new alchemical understanding of how the forces of purpose, creation, and transformation within each of us, when used consciously, can enhance the meaningfulness of everyday life.

In short, my goal with this book is to show that within your own mind and body lies a majestic story filled with drama, pathos, humor, intelligence, fantasy, and fact. It is no less than the story of the entire universe, particularly its own creation, transformation, and ultimate purpose. And while most stories require a separated listener and a storyteller, in your story the listener and the storyteller are one. Here you will see that the way in which you go about telling a story to yourself—a story that includes *you*—actually points out that *without you* there wouldn't be a universe! And we shall see how this story called *you* unfolds into a panorama of life, literally a *you-niverse*—our ultimate goal being to understand the sacred transmutation of mind into matter.

A Word about the Chapter Headings

Because of their symbolic meaning, as explained earlier, I have opened each chapter with a Hebrew letter-symbol. Each letter's sacred meaning will, I believe, enrich our understanding of the material within its corresponding chapter, as well as our overall understanding of the new alchemy itself. Briefly:

א **aleph:** the impossible life-death principle, the void out of which everything emanates[7]

ב **bayt:** the first or primordial container, the first act which distinguishes one thing from another

ג **ghimel:** the first or primordial movement, a seed-like spasm or discontinuous jump

ד **dallet:** a doorway and a resistance to passage or movement, the first resistance or property of inertia [8]

ה **hay:** the first life form

ו **vav or waw:** endless fertility or ability to clone endlessly

ז **zayn:** the first possibility, the concept that possibilities can arise

ח **hhayt:** a gathering or pooling of these possibilities

ט **tayt:** the first actual structure that comes from such a gathering

There are twenty-seven Hebrew letter-symbols in all. (The first twenty-two form the standard Hebrew alphabet. The next five repeat five of the original letters but are written in a modified form to signify their use at the ends of words.) They are arranged in three rows of nine letters with the top row containing the first nine letters. The other two rows are taken to be projections of these first nine and consequently have similar meanings. The difference in levels depends on the evolution of the symbol. Thus free spirit aleph (א) evolves into yod (י) trapped spirit or existence, and at the next level into qof (ק) the cosmic aleph where the reconciliation of spirit with its trapped self occurs.

And thus we see that the Adam Kadmon in Hebrew, אדם קדמון, has a symbolic meaning. Reading this name in Hebrew from right to left, the letters are aleph-dallet-mem (Adam), qof-dallet-mem-vav-nun (Kadmon). In brief, aleph meets resistance (dallet and mem) and finally transforms this resistance of consciousness into cosmic possibilities. The resolution of this is the impossible life-death principle merging with its cosmic destiny and quantum leaping through resistance and the existential resistance of consciousness in endless fertilization of cosmically enlightened Human Beings. Thus, the full realization of Adam Kadmon comes from the sacred transformation of mind into matter.

And, a Quick Word about the Chapters

The chapters in *Mind into Matter* each contain a particular thought concerning the overall process of mind's transformation into matter. As such, each chapter can stand alone as, say, an essay. As the ancient mystics knew, this material is difficult for the human mind to comprehend, probably more so for us today when, as a culture, we're so steeped in the mindset of "objective" science. So, I hope you'll take the time to read and reread those chapters which you find the most difficult. I believe the recursive process will ultimately lead to your understanding of the science and spirit of mind's transformation into matter. More importantly, I believe that once you leave the safety of our old, accepted beliefs, you will begin to see yourself and the story of your life from a new perspective.

aleph

א

Aleph represents the supreme energy—subtle, alive, but not existing as itself in the space-time world we know, because it is imaginal. It is the primal energy in all; and all that we know is in aleph.

Aleph is beyond definition, incapable of being defined or limited. It moves at infinite speed and thus evades time. It is primal consciousness unknown to itself. Its action in the temporal is explosive and discontinuous.

CHAPTER 1

Void: The Impossible Life/Death Principle

For behold, the kingdom of heaven is within you.

Luke 17:21

The main idea of the new alchemy, the cord that binds together all of the ideas presented here, lies in the concept of unity: *the great inseparability of all things.* Taken literally, as we shall see, this means that the very notions of heaven being separate from earth, mind separate from body, free will separate from determinism, life separate from death, and in fact all duality, all opposites, wherein we pose an inside and outside, a boundary line, a nation, an island, a membrane, a distinction—all and more, are not primary facts.

Yet, we unconsciously strive to keep this secret buried inside ourselves. We unwittingly work at maintaining the status quo. In other words, we unconsciously choose to live under the illusion that everything is as we see it. This is not only a fundamental truth for you and me, it is the deep secret of the universe's existence: Hide from one's essential self. It is God's great trick; and it only works because we agree to believe the trick. If we can stop believing it for one minute, one second, even one millisecond, and allow our consciousness to become aware that we have stopped, we will see the trick revealed.

At some point in our lives, somehow, somewhere, just for an instant, the unveiling of the great mystery comes to pass. God, the magician, raises the curtain, reveals the trick just slightly, and we glimpse the illusion. But, we don't shout, Wow! No gasps of wonderment fill the theater. Something becomes distinguishable from nothing in a single creative act, but we trick ourselves into not seeing. And so it goes. No applause fills the air. We sit back, watch the show, breathe a sigh of relief, and say unconsciously, "We'll never figure this one out, might as well just accept it."

In fact, all distinctions arise out of such actions. And most of us habitually remain unconscious and cling to the illusion until the last nanosecond of our existence. We watch the boundary between ocean and land, between air, earth, and water. We watch the effervescent crust of sand, water, and air and remember the distinctions. And likewise, we live our lives in the comfortable notion that an invisible membrane separates us from that world "out there"; that "in here," in our minds, our inner worlds of imagination, we are safe and alone. In no way can any person or thing intrude into our individual mind worlds. Every sense in our bodies continually tells us that this is true, that we are each alone. We ignore any information, any thought, any perception, any imaginative tale, anyone else's story that confronts our sensory presentation of the separated "out there" and "in here" worlds. We look skeptically at people who tell us a different story, probably dismissing them as misguided fools, or even lunatics.

Many of us today, caught in this dilemma of illusion and reality, would like to believe that separateness is an illusion. We are, then, in luck!

What the Alchemists Knew

Distinctions are not real. They are fleeting whispers of an all-pervading, subtle, non-expressive potential reality. The world is not made of separate things. Mind is not separate from matter. And you are not separate from any other being, animal, vegetable, living, dead, or seemingly inanimate matter. The kingdom of heaven and the island of hell lie in you. In you lies everything you have always wanted to know. A vast potential urging itself to arise and become something lies in you. In you, like a coiled serpent waiting to spring forth from your deepest shadows, lies every creative moment that exists, has ever been, and will ever be.

But like the ocean washing ashore, the tide eventually wanes. The water returns to the sea. The shore asserts itself. Eventually all distinctions disappear. No boundary lasts forever. Nothing lasts. Everything returns to the great ocean of oneness. Life, death, and all patterns move vibrationally. You can think of this as the *impossible life/death principle.*

Even space and time—the arena in which we spend our lives—are not real but projections coming from something far deeper and mysterious. Even this arena will disappear. This impossible non-spatially extended, non-thinking thought that lasts not a second nor an eternity, not even the smallest iota of time nor the grandest eon, this deepness, this light/darkness beyond anything that can be pictured as empty, this paradoxical life/death principle, this deep yearning appears as a cloud, a memory, a slight perturbation, and, like that, it grows. But to us, it seems to just pop into existence without a thought or notice.

Ever undulating, the great surge asserts itself once again. The ocean washes ashore. It is an illusion brought on by the very necessity that the action that brings the universe into existence requires this illusion.

Or does it? What if the notion of finding the truth itself is purely imaginary?

The Ancient Alchemists Who Sensed the Void

Inseparability is elusive, most times invisible to our senses, and hard to describe. Yet throughout the ancient world many isolated alchemists sensed the presence of this impossible undivided simultaneously existing life/death principle.

Not unlike some deeper-thinking modern scientists of today, who in their discoveries of new principles of inseparability[1] seek the hidden meaning of life and answers to the mysteries of the universe, alchemists sought ways to bridge the apparent gap that every distinction implies. Behind every good, they sought the evil. Behind every new idea, they sought the ancient principle. Once convinced of their vision, they believed that any separation, discovered or sensed, was illusionary. And, thus they sought a way that would lead them into the realm of inseparability. They desired to hold the paradox of existence in their hands. They desired to see both sides of the coin at once. All of their work, all of their experimental efforts had one goal: shatter the membrane of separability. To do this meant not only working at their alchemical craft, but also working on themselves by continually confronting their own comfort zones of acceptability.

Dissolving the Ancient Membrane

Thus ancient alchemy had as much to do with self-mastery as with mastery of the physical laws of Nature. Such mastery required patience, observation, and, above all, devotion. In the fourteenth century, an ancient alchemist, John of Rupescisia, wrote that alchemy is "the secret of the mastery of fixing the sun in our own sky, so that it shines therein and sheds light, and the principle of light, upon our bodies."[2] To discover this secret, alchemists had to learn to master the art of dissolving all barriers of separability. These barriers particularly included any ideas or concepts indicating a sensory distinction between "out there" and "in here." Thus the most significant membrane they had to dissolve was the one separating mind and matter. They sought to make clear to themselves the invalidity of the distinction between the *real* and the *imaginal* worlds.[3] To do this they had to discover how to cross over willfully and consciously from one realm to the other. This was no easy task because of the law of *inertia*.

Figure 1.1. Dissolving the Membrane. The ancient alchemists sought ways to dissolve the barrier between the reality they believed was an illusion and the imaginal they thought was real.

The Secret Law of Resistance: Inertia

For alchemists, inertia not only appeared to pervade the outer world, tending to keep things in their respective separated places, it pervaded the inner world of thought and perception, tending to make them accept as objective fact what could be repeated over and over again. Indeed, Isaac Newton, an alchemist himself, discovered the universal principle of inertia, which led him to formulate the objective mechanical laws of motion.[4]

To make the break, to overcome mental inertia, requires a new way of thinking. With a new way of thinking, new ways to evaluate what one thinks appear. And with these new tools of evaluation, new ways to feel arise. Now by "feeling" I am not referring to the sense of touch; I mean what Carl Jung meant. Feeling means undergoing a lasting assured occurrence, becoming aware of an "in here" experience through time, but not conscious of the time the feeling lasts—becoming conscious, over a period of time, of a specific kind or quality of a physical, mental, or emotional state. The lasting quality of a feeling is extremely important in what follows.

Once a new way to feel about your thoughts arises, you begin to sense the "out there" world with new eyes: creatively, informatively, newly, as a child. With new enlightened sensations arising, you begin to have deeper intuitions. These intuitions arise as ideas, insights, predictions of the future, or reappraisals of the past. They appear as visions. And, each new thought institutes a cycle. The cycle moves through phases, like those of the sun and moon, from thinking to feeling to sensing to intuiting which begins the cycle anew. The continuation of the cycle forms a physically repeating vibrational nervous energy in the body. Without anything to disrupt it, it forms a memory that could be retapped as if one were going to a beer keg for a refill.

In this cyclic manner all memories form, all impressions become stabilized as "facts," all opinions about the world and opinions about yourself in that world form. When the cycle breaks, when the addictive habit of it dissolves, a new cycle begins. A completed cycle, you see, has what it takes to become a part of reality: it has inertia, it has resistance, and it will, if fueled with energetic cycles that are in phase with it, grow and live. If it grows and lives without check it becomes an archetype and it possesses the user as assuredly as any demon possesses a medieval philosopher hell-bound to tap the secrets of God.

One such secret was continually revealed to the ancient alchemists and it possessed them. It appeared to them in dreams or it arose in their thoughts as they monkeyed around with matter in their laboratories. They were given an inkling that what happened here on earth (the lower world) was linked to what happened in the heavens (the upper world), and that what took place in the inner world of the psyche transforms the outer world of stars, people, places, and things. They had seen how information could transform into matter. And they had seen the reverse. They had tapped the life-death principle. They had reached into the void and dissolved through the membrane of inertia.

Ancient Mysticism and Modern Science

We might think of these ancients as misguided. Perhaps they were, in that they didn't know modern science. But in regard to the essential life/death principle they were, scientifically speaking, right on the money. An inner abstract imaginal world had to have a causative effect on the outer material world and vice versa. As above, so below. As within, so without. The sovereign states of the imaginal and the real are deeply connected.

Long suspected but often doubted and buried, this fact of nature reemerges in the science-based era we call the information age. It leads to a new vision of reality—one where the imaginal, subjective, or virtual reality of mind and the physical, exterior, or objective reality of matter link inextricably. This link, as we shall soon see, transcends time and space.

Because this link lies outside the spatial-temporal realm, many intelligent people believe the answer to the puzzle relating the outer world of substance and the inner world of knowledge does not lie within the world of science, but must be solely found in the metaphysical world of spirituality. However, as we shall see here, these two worlds are joined together outside of space and time into a single worldview as intimately as space and time themselves marry in the Einstein-Minkowski theory of relativity. They are as deeply linked as mind and matter, as real and imaginal. The mind is not in the brain; the brain is not in the mind. In a sense, they can be seen as bordering countries, or as global areas or hemispheres. From separate perspectives, each can be viewed as being contained by the other. From outside they appear as a unity.

Even today we have a difficult time believing that the imaginal world has a causative effect on the real world. On the other hand, we have little difficulty believing that the material world influences the mental. After all, we know that mind-altering drugs can change moods, make pain seemingly evaporate, and even allow diseases to be diminished. Medication can transform a schizophrenic sufferer's brain so that it delivers an apparently normally functioning mind.

The Lunacy of the Alchemists

Today, the boundary separating the insane from the ingenious is not very distinct. Being thought "crazy" was no less the plight of the ancient alchemists. An aura of lunacy shone on all that the alchemists touched. Indeed, to many who witnessed their actions, to be an alchemist was to be crazy.[5] This is because to accomplish their seeming miracles, alchemists needed to go to places in the mind that few dared to reach. And there, according to legend, they had to face a deceiver, a trickster, who stood at the moonlit border between reality and imagination.

But, how did ancient alchemists accomplish their seeming magic? How did they cross over the boundary and evade the ever-present trickster at the border? Metaphorically, they had to be guided by moonlight, no doubt. And they had to expect to see the trickster who, once they recognized him as an image of themselves, would let them pass.

Their moonlit worldview, their whole way of seeing the world, took it that:

Heaven above
Heaven below
Stars above
Stars below
All that is above
Also is below
Grasp this
And rejoice.

Alchemists saw the upper and lower worlds as analogies to the inner (personal) and outer (external) worlds, what we today call the subjective and the objective. In our new alchemy, these worlds are connected by multiple story lines, possibility histories related to the quantum physical idea of *paths of action.* (I'll explain this more fully in chapters 7 and 8.) It turns out that only if we risk facing the trickster at every point along the path can we alter these story lines.

As you move along your own story line, mind-objects—the contents of a virtual reality within the subjective realm which often appear in your dreams as dark characters—take on life and appear to you as new images, thoughts, feelings, and intuitions. You can be swept away by these images as if you were carried by a powerful wave.

The story line connects the "out there" world with the "in here" world. The wave of life moves the self from the mind-objects of the story line into the physical realm where it animates material counterparts. Then the material counterparts react and send back along these same story lines an echo wave establishing a connection between the inner virtual reality and the outer physical domain. This imaginal wave initiation/physical echo wave response results in a loop in time—where the physical activity occurs before *or* after the mind-object appears. When the physical activity occurs after, you experience it as wish fulfillment. When it occurs before, you see it as *déjà vu* or you have an inner sense of knowing what is about to happen.

Remember the first being, Adam Kadmon. From his eyes, mouth, nostrils, and ears, unconfined primal light emanated. In a great overwhelming mystery, special vessels containing this primal light appeared out of nothing. These vessels were primal or seed-like matter. They were the first acts of limitation.

To become this Adam, you need to realize this mystery of light containment. You must tap the time story line, go to the imaginal well, and take the first step to magic and to a sacred awakening. This step was and is the beginning of *something* rather than *nothing*. It was and is the primal act of creation. It was and is the formation of the most powerful tool ever created: the word.

bayt

ב

Bayt represents any container, any physical support, any gestalt. It is the first or primary divisor or separator, for to contain or hold is to separate that which is held from that which is not. It is the primary act, therefore, of consciousness recognizing itself.

If aleph is spiritual, bayt is material.

Chapter 2

The Word: Something from Nothing

I think the universe is a message written in a code, a cosmic code, and the scientist's job is to decipher that code.

Heinz Pagels, physicist

In the opening scenes of a popular 1960s television series that played to a select but mystified audience, the hero, known as *The Prisoner*, a nameless fellow who has suddenly quit his top-secret government job in England, is kidnapped from his London flat and interrogated by an antagonistic and mysterious inquisitor.

"Where am I?" The prisoner asks.

"You are here," replies the inquisitor.

"Where is 'here'?" continues the prisoner.

"Never mind that," says the inquisitor, "I am number two, you are number six."

"I am not a number," pleads the prisoner. "What do you want?"

"We want information," comes the inquisitor's reply.

After the questioning and imprisonment in a mysterious village whose location is unknown, the prisoner embarks on a number of harrowing adventures, each calculated to determine whether or not he will bend to the rules of the mysterious organization to which he has been made captive. The goal of this secret club is to force him to yield some precious knowledge, which the prisoner seems not to know, to make him a cog in the giant machinery that carries out whatever mischief the occult conglomeration chooses to perform. Our hero resists, of course, but at the price of his own sanity.

In a sense, we are like the prisoner in the story. We live in "the information age." Facts and data seemingly jump—metaphorically and factually—quantum levels, impacting all of us. In our Internet-web-connected world, while no one would doubt the amount of information "out there," few would consider how it "secretly" affects us: information shapes our mental reality, our lives, our bodies, and the material world we inhabit.

Information transforms our everyday reality regardless of our ignorance. It moves and forms our thoughts and words. It makes up our vocabulary. It crosses both language and geographical barriers creating new concepts. It frightens us. It excites us. At times we feel the need to "get away from it all," meaning newspapers, the office, television, and other media. At other times we feel the need to seek out these media to see "what's happening." Information offers us new meanings to old ideas, and it affects the ways in which we conduct our relationships with others and with ourselves—even if we, like the prisoner, have never been privy to the source of that shaping and transforming intelligence.

Information is both a bridge—the medium—and a message, à la Marshall McLuhan's *The Medium is the Massage.*[1] (Yes, the word is "massage," not "message"; McLuhan's play on words.) It connects two seemingly different worlds: the so-called *real* world that we all sense "out there," and a world few of us think about although we experience it every day as quite real—the *imaginal* world. We all experience "in here" the universe of our dreams, hopes, and fantasies, the inner world. Perhaps even more surprising, information (the stuff of the imaginal) not only *transforms* the material world, it *becomes* it. The old adage "you are what you eat" has changed into "you are what you know" and since your knowledge ultimately depends on what information you accept as "fact," you are what you believe!

The Secret of Information: the Word

In the beginning, according to what we presently know about origins, there was nothing, nothing at all.[2] But then, something rather miraculous happened. Matter, antimatter, energy, space, time and, most significantly, *information* suddenly spewed into existence. It was, apparently, a kind of fluke. The universe of *nothing*, with nothing better to do, created *something*. According to our best scientific computation, if we add up all the energy in the universe, including all contained in matter and antimatter, and including the attractive kind in gravity, we come up with a big zero. It all adds up to naught. But if we add up all of the information in the universe, it appears to be nowhere near zero. Indeed it is infinite. And that is where human beings come into the picture.

When that first curtain opened to God's Magic Theater, a great void appeared. And then, according to one myth called Science, the void exploded into the Big Bang. Following another myth called the Bible, in the beginning there was the Word and the Word was with God, and the Word was God.[3] These seemingly very different points of view—these myths called the Big Bang and the Word—appear entirely unreconcilable: one deals with the physical universe of matter and energy and the other with the mental universe of mind and information. But, could these two views actually be saying the same thing? Could it be that in some way what we describe about the universe—how we exploit it to derive meaning from it, how we determine what it is, and what it is doing—establishes the very universe we speak and write about? Does the act of learning something, turning our experiences into *meaningful* symbols of discourse create both the physical thing being observed and the laws of order it seems to obey?

The answers to these and other questions will come from a harmonizing of the relationship between the "in here" world of information, meaning, and knowledge with the "out there" world of matter, energy, and existence. This reconcilement is precisely what I mean by a *new alchemy.*

Sticks and Stones: Bits of Knowledge

Where does such a rapprochement begin? It starts with words and naming things. As a child, I remember, sometimes painfully, the names we kids used to call each other. Although we all knew the jingle, "sticks and stones will break my bones, but names will never hurt me," we often found names did hurt us, sometimes leaving painful wounds and scars that perhaps have never healed. "Dick is a moron"; "Fred is a dirty Jew"; "Your mother is a fat pig." I'm sure if you dig back into your own past you will also recall somewhat similar graphic and painful moments of angry defilement. Words can and do hurt.

In the sixties when the issue of minority rights was becoming a major topic of discussion, many found that even talking about racial issues was offensive. Comedian Lenny Bruce, who was known for his scatological language, deliberately used racial epithets when his audience consisted largely of members of the specific racial minority to whom the derogatory words were hurled. He would repeat an offensive word until it became a meaningless mantra. Then he would explain to his outraged audience: A word is just a sound; everything depends on how the audience, or the individual, reacts to the sound. He explained that when we allow words to hold power over us, when we attach negative or fearful emotional significance to words, we suffer. Thus, by not letting words invoke images, the words lose their power and become meaningless sounds.

Well, that may be easy to say, but as we all know, words can and do leave their marks on us. In her book *Refiguring Life*, Evelyn Fox Keller points out that "the notion that words are one thing, acts another, was radically undermined when linguist J. L. Austin laid out his theory of 'speech-act' in a series of lectures at Harvard University entitled *How to Do Things with Words*."[4] Austin showed that words are not always descriptive but are often action provoking. Examples include marriage vows, declarations of war, or the classic Jimmy Cagney line, "Take that, you dirty rat." Keller goes so far as to state that all language is action provoking and this includes scientific language as well.[5]

But how? How can a word hurt me?

This was not a mystery for the ancient alchemists. They saw the physical world in which we live as being objectively "out there," as only a part of a vast system of other realities connected to the physical world. Being non-physical worlds, we might think scientists would tend to dismiss them. But that would be a serious mistake, as serious as believing that only sticks and stones break bones.

It From Bit?

Quantum physics and modern computers add a new view of how our words—the stuff of our minds—alter and shape the world we all take for granted as "out there." In effect, there is no "out there" out there unless there is first and primarily an "in here" taking action—one having a deep transformative effect—on the world "out there." *No reality without a perception of reality* is the first rule of quantum physics according to visionary physicist John A. Wheeler. From this viewpoint, an object is whatever your *six* senses (seeing, touching, tasting, hearing, smelling, and "minding," using the Buddhist notion of mind as a sense) tell you it is.

The description of an object defines the border between objective and subjective reality. At least half of my experience appears to be in the indescribable inner world of my mind. We are each privy to this world as if no other were present. The new alchemy brings this simple assumption into question. From a new alchemical point of view the unknown territory of my mind—to which I am apparently privy and not you—is the same as the mind you enter. We each experience one mind.

Herein lies a secret to alchemical transformation. This vast land of the imaginal is not yours and yours alone, but the same territory that each and every sentient life form enters—even life forms that do not speak. As you wander this inner kingdom of one mind, you find the other voyagers to be invisible.

This inner world, sought for by the alchemists of old, may now be understood using a new and perhaps universal language. This is the language of the "bit," a single smidgen of information, a tiny switch that points upward or downward and indicates a simple "on" or "off," a simple single digit of one or zero, all or nothing at all. Let me explain a bit about bits.

The notion of bits comes from the vernacular used to describe information in the modern computer. Bits are data of the simplest kind. A bit has either the value zero or one. That's it. Strings of bits like *0-0-1-1-1-0,* and so on, can represent numbers, codes, instructions, pictures, moving objects, three-dimensional virtual realities, and if some scientists are correct, even reality itself.

One can think of any computer as a hole-punching and repairing device that has three very long reels of paper tape running into it. Each tape is divided into equally sized squares running the length of the tape. Each square contains a single bit indicated by the paper's being punched through, or not. The first reel, *the data*, runs into the machine through a slot. These data are then read by a light sensor that detects if the square on the paper tape has been punched through or not.

The second reel, *the program*, runs through another slot and contains instructions telling the machine what to do with the data. These instructions are themselves merely bits running sequentially along the tape. For example, if a sequence of four bits on the program tape contains a string of *1-1-1-1*, the computer could simply follow the instruction: "Read tape 1. If you read the bit *1*, punch a hole in tape 3 and advance tape 3 one square; if you read the bit *0*, do not punch a hole in tape 3 and advance tape 3 one square." If tape 2 had a different string of four bits, say *1-0-0-1*, the computer would follow a different set of instructions. For example: "Read tape 1. If you read a *1*, advance tape 2 three squares; if you read a *0*, retract tape 2 four squares."

The third reel, *the result* (or *output* in the jargon), is outputted through a third slot. Its squares are normally blank and only punched, repaired, or left as they are by the machine when it "reads" the data on reel 1, and does something with it according to the instructions on reel 2. At the end of the computation, the output tape contains a sequence of new information generated by the raw data of tape 1 and the program of tape 2.

The amazing thing about this simple "computer" is that it describes all possible computers.[6] The punched holes or their absences on all of the tapes simply represent *0*s or *1*s, ons or offs, or dots or dashes. Amazingly, this most basic set of information consisting of a string of single *bits* enables airplanes to receive instructions on how to fly, automobiles how to drive, and people, if the new biology is correct, how to live.

The bit, the smallest and simplest thing we can imagine, makes the first distinction. It shows how one thing is distinct from anything else. In modern computer technology it is universally recognized as the fundamental unit of this vast system of knowledge. As far as we presently understand, anything objective is reducible to bits of information arranged in sequences. Arrange them in time and we have the telephone, the radio, Stravinsky's *Rites of Spring*, the Beatles' "I Am the Walrus," and every thought or feeling that has ever or will ever occur. Arrange them in space and we have Rembrandt's *The Night Watch*, Picasso's *Guernica*, the distance to the moon, and photographs of the surface of Mars. Arrange them in time and space and we have modern television, Ingrid Bergman in *Casablanca*, and perhaps, with the advent of virtual reality technology, the appearances of three-dimensional imaginal worlds that we each can enter.

Here we see that bits are the basis for all distinctions in computers. In neurophysiology one might guess that nervous systems and brains sense bits. Bits transform into experiences from the highest to the lowest, from the simplest to the most complex. Bits arranged in space and time become information "out there" in the objective world. Experienced as thoughts and feelings, they are the information of our "in here" imaginal world. A sequence of single bits creates the distinction between the "in here" and the "out there." For without information, how do we know what's "out there" as distinguished from what's "in here"?

Who Wrote the Program?

Physicist John A. Wheeler calls the information basis for this new alchemy *it from bit.*[7] "It" from "bit" refers to how a thing, an "it," arises from a single "yes" or "no" answer—a "bit" of information. Sounds simple, yet I find something hidden in the above description. If everything is made from bits, how did the computer that senses those bits get built in the first place? For example, what gave it the instruction to recognize that a sequence of four bits on tape 2 (program) meant that a single bit on tape 1 (input) should be read and a single action involving tape 3 (output) and tape 2 (program) was to be taken?

In modern computer parlance this is known as the bootstrap problem. The notion of bootstrap comes from the saying "hoisting oneself up by one's own bootstraps." Every time you start up your computer, it follows a basic set of instructions that are, so to speak, hardwired, or bootstrapped, within the computer. These instructions must "come alive" when you turn the switch that allows electricity to come into your machine. The computer must know that it is to read the input tape for example.[8] The bootstrap program essentially makes the computer ready to learn and perform its functions. In essence the computer "knows" what it is to do with programs, how to "input" information, and how to "output" results.

Imagine that you are God and you are up on a mountain somewhere with all of time, space, matter, and life before you. Now suppose you want to create a universe of intelligent life. You want that life to feel as if something matters, something in life is important, something is more important than something else is. In other words, the life you are creating must have purpose.

So, you imagine life forms that have the ability to learn and evolve in time, to grow and become more intelligent. But how are you going to accomplish this miracle? How are you going to engender life and give it the ability to grow and learn?

Well, I don't know how to create life from nothing, but allow me to hazard a guess as to how I might do it if I were God. I would need to set up just the right bootstrap instructions. These instructions would need to be intelligible enough that, once started, simple life forms could learn, in essence, how to read a program, how to read data fed to programs, and how to output the results of those programs to itself and other life forms around it. The program would also have to be simple enough that the simplest life forms I could imagine could follow them. Today such programs might be called instincts, but it's difficult to regard all life forms as having the same instincts.

Or is it?

ghimel

ג

Ghimel represents movement—the motion of all bayts (matter) containing aleph (spirit). In order for movement to exist, space-time is necessary; so ghimel may be viewed as the primary seed of space-time. And ghimel is not possible without aleph and bayt.

Aleph, bayt, and ghimel are the primary seeds from which matter-space-time, the so-called arena of existence, is manifested.

Chapter 3

The Mind in Body: The Desire to Move

Your mind is in every cell of your body.

Candace Pert, neurobiologist

Perhaps our primary instinct as human beings is to make the world distinct—to put distance between things. First mentioned by Empedocles (450 B.C.E.), the ancient Greek alchemists (others might call them philosophers) distinguished earth, air, fire, and water as primary elements in objects. They asserted that each thing in the universe is made from different combinations of these fundamentals. When the alchemists practiced their art, they transformed, or freed, the elements from their compositions, returning them to their natural places in the universe.

Under the guiding hand of the alchemist, elements could be recombined into new forms of matter by simply moving them to new locations or by changing their qualities. By recognizing the ancient laws of movement, new forms could be created.

One hundred years or so after Empedocles, Aristotle asserted that each of the primary elements is composed of a mixture of two qualities taken from a list of four: hot, dry, moist, and cold. Thus fire is dry and hot; air is hot and moist; water is moist and cold; and earth is cold and dry. To transform a substance, then, we would need to change the qualities in the elements making up the substance. For example, by driving out moisture from water the opposite quality, *dry*, would emerge, turning water into ice. By adding the quality *hot* to ice, *cold* is driven out, turning ice into the substance that is *hot* and *moist*, namely vapor, or air. By driving moisture out of air we are left with something that is *dry* and *hot*, namely fire.

Every material object is a mixture of different amounts of the primary elements of fire, air, water, and earth. Left to their own devices, objects moved according to the fundamental law: each element to its natural place. Thus when an object decomposed, it would break up into its natural elements and these, in turn, would move accordingly.

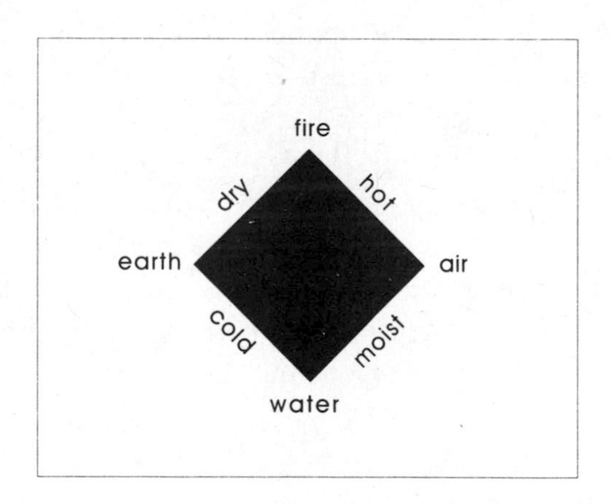

Figure 3.1. The Ancient Greek Alchemy.

Fire, being the lightest element, naturally sought its way to the heavens above. Air, being heavier than fire but lighter than water, would rise above water and earth, but not reach as high as fire, which was imagined to reach the stars. Water sought its natural place below air but floating upon the land. And earth, the last material element, was the foundation for all material substance. This was the natural law.

Missing from Aristotle's explanation, however, was the fifth essence, the quintessential spirit that enabled individuals to resist being completely engulfed by the ever-persistent deterioration into the first four elements. Somehow, in opposition to the natural laws of decay, living beings formed and were able to move, resisting the unconstrained motion following the usual law of decomposition that eventually awaited them all at death. This motion was archetypal, fundamental, and vital.

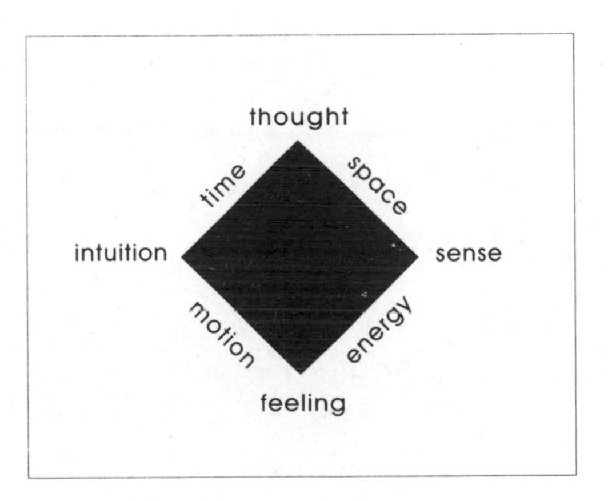

Figure 3.2. The New Alchemy.

Living creatures used something to overcome the natural law. But what was this mysterious essence? The old alchemists knew it as the *quintessence*—the fifth essence—of life. The new alchemists know it as *spirit* with its quality motivated by *desire*.

The Four Axioms of Desire

From the point of view of the new alchemist, desire provides the primary basis for all transformation. To evolve we must desire. The rules of desire control everything. And, to work with desire, we need to know the four axioms of desire:

1. If we think, we become time (thinking → time).
2. If we sense, we become space (sensing → space).
3. If we feel, we become energy (feeling → energy).
4. If we intuit, we become motion (intuiting → motion).

Now let's look at these four axioms individually.

Axiom 1: Thinking about Thinking

Axiom 1 refers to the activity of thinking. When we think about something, just absently, not really concentrating on it, our sense of time disappears. As we pour our desire into thought, we become time.

Becoming time is not difficult. It is like entering a flowing stream and allowing ourselves to drift along with it. As we flow with the river, the water that appeared to be rushing past when we stood upon the shore suddenly becomes quiet and motionless. It is the same when we enter the river of thought. We lose time simply because we become it.

The theory of relativity tells us that time is not absolute. Fixed time intervals, say one-second units, are not equal for a moving clock and a clock at rest. As a thing moves faster and approaches the speed of light, a one-second time interval stretches, covering longer and longer time intervals as determined by the clock at rest. At the speed of light, time slows so much that it completely stops. Or, to put it another way, any time interval, no matter how short, stretches to infinite time measured by a resting clock. And so it is when we enter thought.

This is the relativity of time. Time is sensed in beats. The ability to "sense time" is fundamental to human experience. When we sense time it means we compare *a new sequence of experiences* (NSE), new beats, with an already *mapped sequence of experiences* (MSE), old beats. Usually the MSE are well known as, for example, observing your watch from time to time, or the simple ticking of a clock, or the clicketty-clack of wheels on rails when riding on a train. If the NSE consists of mind wanderings, with few thoughts about what we are thinking about, and we then compare this NSE with the MSE, it feels as if time passes very quickly. Below we see twelve MSE (🕐 events) taking about the same amount of time as six NSE (☺ events).

MSE 🕐 🕐 🕐 🕐 🕐 🕐 🕐 🕐 🕐 🕐 🕐 🕐 (12 🕐 events)
NSE ☺ ☺ ☺ ☺ ☺ ☺ (6 ☺ events)

On the other hand, if you have a sequence of rapid "worrying thoughts," *then thinking about thinking* (TAT) events occur. When many TAT events occur, and you then compare this TAT sequence with the same MSE, it feels as if time passes very slowly. For example, see the picture below. It indicates time passing slowly for the TAT thinker.

MSE 🕐 🕐 🕐 🕐 🕐 🕐 🕐 🕐 🕐 🕐 🕐 🕐 (12 🕐 events)
TAT ☹☹☹☹☹☹☹☹☹☹☹☹☹☹☹☹☹☹☹☹☹☹☹☹☹☹ (26 ☹ events)

We always compare TAT experiences with other experiences, usually MSE, to determine "time." Thus, the subjective sense of time, without reference to any clock or MSE, would be the same for the following two TAT NSE:

MSE	🕘 🕘 🕘 🕘 🕘 🕘 🕘 🕘 🕘 🕘 🕘 🕘	(12 🕘 events)
NSE 1	☺ ☺ ☺ ☺ ☺ ☺	(6 ☺ events)
NSE 2	☹☹☹☹☹☹	(6 ☹ events)

When we compare NSE 1 with NSE 2, since they each consist of six TAT events, they would appear to us as if they took the same amount of subjective time. But, when we compare NSE 1 with the above reference MSE, we see NSE 1 as time passing quickly, since twelve MSE events or ticks of the clock occurred as compared to six NSE 1 observations of TAT. Similarly when you compare NSE 2 with MSE, you see NSE 2 as time passing slowly since six NSE 2 events occurred as compared to perhaps three ticks of the clock as seen in MSE.

This reflects what I call the complementary ways in which you can observe your own thoughts.[1] From quantum physics we know we can't observe both position and momentum of any object simultaneously. It appears that the same rule also applies to our thought processes. Presumably, T (thinking) requires the movement of particles (perhaps electrons) in the brain and nervous system, while TAT requires the comparison of one T experience with another. So T and TAT cannot be done simultaneously; they are complementary to each other. A T experience arises from observing positions of electrons in some manner, while a TAT experience arises similarly from observing momentum of electrons. With TAT experiences we compare one T experience with another T experience at a later time. The comparison of one T event with a previous one must occur for a sense of flow or momentum to arise.

Thus, that we experience no sense of time when we are actively engaged in thinking tells us that time and thinking are the same—they measure the same quality of the universe. This is akin to stepping into the flowing river of time and drifting along with it. While we drift we don't sense the movement of the stream. Only when we leave the stream or fight against it—think about what we are thinking about—do we feel it. If we externalize that quality, we call it time; if we internalize it we call it thought.

Axiom 2: Thinking About Sensing

As with the word "feelings," when I refer to "sensations" I am referring to the way in which Carl Jung used the term.[2] Sensations involve movements of electrons or other electrically charged particles from one place to another. A sensation implies the existence of a disturbing event or factor, such as the prick of a pin on the skin or a molecule of sugar landing on a taste bud. Sensations include vibration, heat, cold, taste, smell, sight, and sound. For a sensation to arise, some location in the body must register it. The skin, for example, registers the location of pinpricks or heat; the tongue registers the sensation of taste. Sensations correspond to the quantum physical operation of locating some particle at some registering device in the body, usually a nerve ending. Thus, to sense anything at all we must become aware of *spatial extent.* Our senses tell us that we have a body. Our body is a space-sensing mechanism.

Swami Yogananda, in describing the various links between normal mental modifications and functions in the Sankhya and yoga systems, helps us understand this space-sensing mechanism: "The different sensory stimuli to which we react, tactual, visual, gustatory, auditory, and olfactory, are produced by vibratory variations in electrons and protons."[3] And, the disincarnate entity Seth, in Jane Roberts's book *The Unknown Reality,*[4] describes the ego as specialized in expansions of space and its manipulations. The ego arose in tribal environments as a necessary specialization; it enabled data from the senses to be differentiated emotionally and *otherwise.* Tribes formed in which members were considered to be either inside or outside the tribe. This tribal consciousness was the first group ego. Later, consciousness was not able to handle the tribal ego as it was, and individuation began to take place. This process depended on cooperation between the members of the tribe. Thus, the individual ego arose as an agreement between the tribal members.

Axiom 3: Thinking About Feeling

Many physicists believe that all matter is ultimately composed of trapped light. This belief is embodied in Einstein's famous $E=mc^2$ energy/mass equation. The notion here arises from the fact that every particle of matter has a mirror-image particle of antimatter. When a particle of matter interacts with its mirror-image particle, it undergoes a process called matter/antimatter annihilation, the result of which is the production of light or massless energy. Thus, matter is trapped light.

In my earlier book *Star Wave,*[5] I speculated that human feelings such as love and hate could be described in terms of simple, primitive base feelings. These base feelings are found in the matter-light energy transformations of electrons and photons—light particles. Hate, for example, is explained as a quantum statistical property of electrons—and no two electrons will ever exist in the same quantum state. While love, on the other hand, is explained in terms of the quantum statistical behavior of photons—and all photons tend to move into the same state if given the chance. Thus, in a physical sense, the phrase "light is love" is more than metaphor.

We could say, then, that the reason we all suffer from loneliness and other types of human pain connected with our material bodies is due to the hate/isolational tendencies/properties of electrons. These electrons are, in a certain sense, trapped light. I speculate that electrons "feel" some form of quantum suffering—a desire to become light once again. When an electron meets with its antimatter partner, the positron, the two particles annihilate each other, producing the extremely high-frequency light known as gamma rays.

I believe that all of our human feelings and emotions are rooted in these simple physical properties of matter and energy transformation, and that human feelings can be explained by looking at the group properties of many electrons in the human body.

Each electron in each atom of matter can be thought to possess well-defined classical properties, such as electrical charge, mass, spin, magnetic moment, inertia, and location in space. The last attribute is, however, suspect. This suspicion is due to the wave-particle duality of all matter—the quantum nature of the physical world. Electrons can be imagined best as "events with attributes," rather than as objects with properties. The electron, in other words, is a construct of human thought. Since human thought is limited to immediate sense impressions and since quantum physics is based upon a world beyond such impressions, none of us can know what an electron really is. I take this as a necessary and sufficient condition for the existence of the physical world: *The physical world exists simply because mind cannot ever completely know it.*

Axiom 4: Thinking about Intuiting

You have a natural intuition. As Jung put it:

> Things have a past and they have a future. They come from somewhere, they go to somewhere, and you cannot see where they came from and you cannot know where they go to, but you get what the Americans call a hunch.[6]

Put simply, intuition reflects your ability to know where you are going and what you are going to do next. If you did not have this natural sense, the ability to intuit, you could not walk a single step, drive a car, or for that matter speak a sentence. Intuition is your ability to sense motion, and motion is the key to change and transformation. Motion, when it refers to spatial change, is reflected in our ability to see change as a difference from that which is not changed. We measure this with our sense of wave motion and our measure of the length of a wave.

Intuition also arises from desire. But—and this is key—intuition is not realized unless there is some change in thought. A shift must occur.

"Out There" and "In Here"

From a physical materialist panorama, the stuff of the world alters and shapes the mind we all take for granted as "in here." This observational *solar* aspect of modern science says there exists a reality *independent* of how or what anyone or thing thinks about it. This reality evolves without any help from anyone to guide it along its Darwinian probabilistic destiny. Indeed, according to this viewpoint, the mind itself is an epiphenomenon arising from electrical activity within carbon-based cellular life forms. Since mind arises from matter it has no power, no ability to change matter. All apparent willful changes that people make in their lives are merely mirages; following Darwin's laws of evolution, the real power for transformation comes from within matter itself.

Accordingly, matter and energy are destined to do what they do. A person has as much power to change that fate, as he or she has to prevent the sun from shining. What rules it all is blind, deaf, and dumb destiny. If we want to call this "God," we may, but here God is an absentee landlord at best, and we are living in illusion if we believe we can change anything. Our actions are the whims of a life form, following patterns of electrical activity set in motion millions of years ago, when we evolved from the great stormy sea where no human being even existed. Intelligence is an illusion. Mind is a convenient label to attach to invisible processes that we are slowly coming to understand in terms of objective processes.

Why does this viewpoint have such a following of believers? Because we can observe objective cause-and-effect forces in motion and we can map these in our minds. But although it is totally missing from our modern scientific viewpoint, in spite of its obvious necessity, the truth of the alchemical viewpoint is becoming apparent to the larger society. Today many philosophers and psychologists feel that the alchemy of the ancients is a metaphor for the transformation of a human being into something new and more spiritual. Jungian psychologist Edward F. Edinger points out that Jung's insights into the inner depth of the psyche establishes its existence, without doubt, as a series of objective facts constituting what he, Edinger, calls the *anatomy of the psyche.*[7] This anatomy is constructed from archetypal images, which act as alchemical devices and speed human evolution on its course.

At the same time philosophers and psychologists are coming to these new conclusions, quantum physics and modern computers have added a new view of how our words—the stuff of our minds—alter and shape the world we all take for granted as "out there." In this view, there is no "out there" unless first and primarily there is an "in here" taking action—a deeper, transformative effect—on the world "out there."

In chapter 2, I pointed out the first rule of quantum physics: *No reality without perception of reality.* I also explained that an object is whatever the *six* senses—seeing, touching, tasting, hearing, smelling, and minding, using the Buddhist notion of sense—tell us it is. To this rule, I add: No *new* reality without a *new* perception of reality. Here we examine how to lay siege to the inner images found in those story lines mentioned in chapter 1. Like the ancient alchemists, we need our "lunacy" to become enlightened, to see things as creation. There is no such thing as an absolute, ultimate reality. There can't be, because each person must perceive "reality," and perception is always affected by what each person brings to bear in the act of perception. This is the observational, *lunar* aspect of quantum physics.

But how are we to map this inner landscape so singularly witnessed? Perhaps this is a fool's folly and there will never be a description of its secret hiding places. Perhaps a trickster standing at its boundary is there to tell us this. Or perhaps we need to make a new effort at naming things invisible—those things behind the magician's curtain. We need to show the trickster a few new tricks.

I suggest that it is possible that the search for the *prima materia* is finally over; that within our midst lie the means to transmute ordinary and mortal life into a new vision of immortality and spirituality everlasting; and that the answer to transformation lies within our psyches—our thoughts and feelings, particularly as to how they deal with the objective world we believe in so fervently as "out there." We have the means to transmute our lives through a new understanding of information, and through how we use that information.

dallet

ד

Dallet represents resistance or response to movement. It appears as the interaction between material objects in states of motion. It responds to all ghimel (movement) of the bayts (matter) containing aleph (spirit).

Dallet appears as the common inertia of materials and it plays a vital role in the universe. It is the recorded response or resistance that is continually offered to ghimel (motion). The prefix re*—as in words like recall, remember, resist, or response—is pure dallet.*

CHAPTER 4

Inertia: The Mysterious Resistance

One step forward, two steps back.

Vladimir Lenin
on the Russian Revolution

"I think, therefore I am." Or is it, "I am, therefore, I think?" Or can we argue that a universe cannot exist without being and thinking—matter and mind intimately tied together? Certainly we know that mind influences matter. It happens every time we lift a finger—we think of it as our minds touching matter. Each of us believes the presence of a single mind inside of our own head.

Yet, surprisingly, there is no scientific evidence that mind exists in heads or otherwise! Can we scientifically demonstrate its existence? Can we go in search of it? It would seem not, as science deals with objective evidence, and mind hasn't a thread of objective existence.[1]

Perhaps we could move closer to proving the existence of mind by asking: Does matter influence mind? Well, we can observe *matter* influencing *matter* when a dropped egg hits the floor. Certainly the floor drastically affects the egg. And when we take a mind-altering substance such as alcohol or a pain-numbing drug, we feel its effect on our ability to be alert or feel pain. So, we tend to believe that *matter* influences *mind.* We could say, therefore, that mind and matter exert a force on one another. But since mind cannot be scientifically discovered, this force cannot be found strictly via the fields of physics or psychology. No, we need a new field, the field of new alchemy, which incorporates the overlapping insights gained from both physics and psychology; which seeks to reveal the secret connection between matter and mind—seeks to reveal the force that we intuitively believe real.

I realize this alchemical force in nature through the most basic recognition that *I*, myself, exist at a boundary between two seemingly distinct regions within the known universe: the *real* and the *imaginal* realms. When I write *I*, I mean the instinctive sense that you and I both have that we are individual beings, that we exist separately, and that we see ourselves as distinct human individuals with diverse and seemingly specific, airtight, conscious minds.

While we sense this individuality in ourselves, we assume without question that others also sense it. Hence we go through our lives sensing that each of us is alone, somehow locked inside of a body, and yet completely able to realize that another also exists with this same sense of bounded existence. Buddhists refer to this isolation as *dukkha*, or suffering. We experience a sense of lack, loss, or withdrawal from the world in some way.

Some of us, from time to time, suspect that others are able to transcend this lonely boundary. Perhaps a few of us have also felt, if only momentarily, that we may have actually done so ourselves. But most of us feel quite separate and alone well within the skins that define our bodily boundaries. Most of us feel this way nearly all of our waking lives.

In describing this dichotomy one might also use words such as body and mind, objective and subjective, outer and inner, and so on. But no matter how one defines these realms, they must be dealt with simply because each of us is singly aware of the dichotomy. The continual awareness of one realm—the real, or bodily—at the expense of the other—the imaginal, or mental—usually leads to some form of disaster in a person's life. For example, to be completely aware of your body while having no awareness of your mind would make you narcissistic. On the other hand, to ignore your body and pay attention to your mind alone would have an equally dire but opposite effect. You could ignore obvious signs of bodily injury, for example, and dismiss profuse bleeding or a broken bone. You might also have little or no feeling for others, viewing them as if they were merely mechanical devices or intelligent robots.

Yet, awareness of both of these realms, real and imaginal, mind and body, is not enough to become successful in controlling them. We must learn to move our intent, at our command, between the two realms. We must recognize the force of mind on matter and also recognize how the force of matter on mind arises. This is somewhat like a form of tai chi or possibly the recognition of how *ch'i* arises.[2]

This is the first secret of alchemical creativity: Learn to become aware of the mind/body split and the point at which our unconscious mind influences our body. To do this, we need to become aware each time we feel resistance arising within ourselves.

"Resistance is futile," say the robotic cybernetic species called the "Borg" of TV's *Star Trek* series. The Borg are a collective hive—a species made of half machine and half human life forms. The Borg tell their captives not to resist and to join the collective. If they don't, they are forced to join anyway. We might think from this that resistance to force of will is a good thing. We would be right. Without resistance nothing comes into existence. Resistance ties mind to matter in spite of science's materialistic attempt to prove mind nonexistent. A good teacher uses resistance to an advantage when dealing with a difficult student. The teacher attempts to bring the student to the point where resistance to a new concept arises in the student's mind. Then the teacher deals with the resistance by implementing the "one step forward, two steps back" technique—a phrase coined by Lenin when describing the 1917 Russian revolution.

When the universe becomes aware of arising resistance—the same resistance you and I experience when we discover a new idea—the universe comes into being and becomes self-aware. It transforms information into matter, giving rise to physical objects, and transforms matter into information, giving rise to a mental universe that models the physical. Resistance arises in each transformation. This is the transformational process. Like a snake looping back on itself, seeking its tail, resistance is a self-referencing, feedback and action/reaction process. It can also be seen as a double movement—an out-breath and an in-breath, an expansion and a contraction. As we take a breath, we experience resistance to more breath being taken in. As we expel a breath, we experience resistance to more breath being expelled.

The first step forward in the process occurs according to a movement in time from present to the future. This is usually called causality or determinism. When we imagine a scene in the future, we exercise the forward step. The second step in the process occurs according to a flow in time from the future back to the present. This is called fate, teleology, or purpose. When we confront our own inertia and get up and move in desire to accomplish the vision seen in the forward step, we exercise the backward step. We might look at this in terms of the moving hand drawing the image of a moving hand. Hand 1 draws the image of hand 2 which draws the image of hand 1, which arises from the image 2 and continues by drawing hand 1, and so on. Artist Maurits Escher caught this vision in the following drawing.

This two-step dance in time seems simple enough, even if a bit mysterious. But just how does this work? And perhaps even more importantly, why does it occur at all and why is it important for each of us to know this?

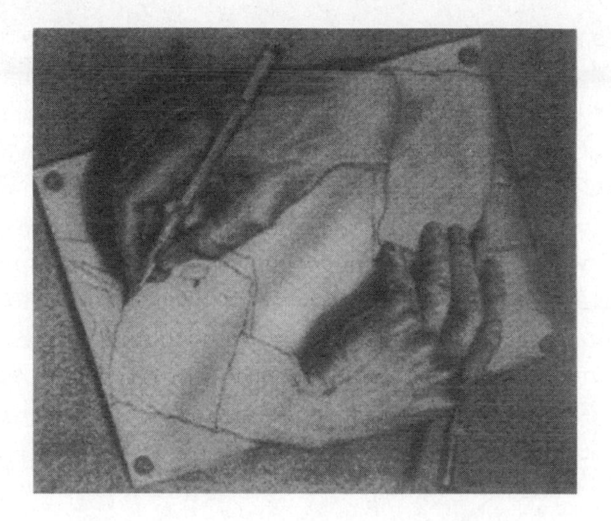

Figure 4.1 Information Shaping Reality Shaping Information. In the beginning was the word. As illustrated in this drawing by Escher, information and matter move continually in a self-referencing interaction that forms instruction to continually transform matter, which in turn feeds back information, recreating matter all over again. The question is, how is this accomplished?

The Trickster at the Inertial Door Separating the Real and Imaginal Realms

Creation is a two-step dance of transformation: It is one hand shaking the other, one hand drawing the other. And transformation is never an easy process. Inertia, a requirement for existence, a property of all material objects, a necessity for even thought to arise, sets itself as the barrier blocking creativity, sapping all new energies that attempt to defeat it. Therefore, if we wish to create we must overcome inertia. As Lenin put, one step forward, two steps back. To create, we must push one step forward from the real to the imaginal realm. The rigidity of our own thinking will result in more inertia than we can imagine—

two steps back. The trickster who guards the gate between the real and the imaginal realms whispers, "Don't pay attention to the inertia," and in so doing makes us aware of it. Hence to overcome inertia we must release previous ways of thinking, not by opposing them, but by allowing them to be expressed. There is no inertia greater than our own previous mind set. Don't be disturbed by this, however, for without this inertia no universe is possible, no matter would ever come into existence, and we would never have learned to talk, walk, or ride a bicycle.

Crossing from one realm to the other is not easy. We need the four functional tools (described in the previous chapter as the four axioms of desire) to navigate through space and time—feelings, thoughts, senses, and intuitions. (Jung called them functions.[3]) We navigate our way through life using them. We bring them to bear when we negotiate the territory of our lives—our lives in day-to-day action. Thus they enable us to move at will from mind to matter and back again.

And the mind maps that we have created for ourselves—our preconceived notions of reality—tell us just which, when, and for how long any or all of these tools are to be employed. However, as much as we use our feelings, thoughts, senses, and intuitions, we can be tricked by inappropriate use of them. In other words, the very tools that we use to take us through our lives are themselves conscious and alive and are capable of acting back on us. They can fool us. This deception occurs when we use these tools to navigate our maps (our preconceived notions of reality) and not our territories (our actual experiences of reality). This cannot be overstated: We misuse our tools when we resist using them to deal with our real experiences in favor of our imaginal experiences.

Let me give you an example. You are mulling over what happened to you during a previous encounter with someone. Perhaps this was a painful encounter, or one that brought fear or doubt into your life. You used your tools, your feelings, thoughts, senses, and intuitions, to evaluate the past situation as it was occurring. Fine. That is indeed what the tools are for.

But now as you sit and mull over just what happened to you, you need to be on guard, for here is when the trickster arises. Your tools speak. Listen to them, but don't become attached to what they say.

We could say that the trickster, with map in hand, stands at the mind/matter doorway and beckons us to cross from one side to the other. We cannot always trust him, however. At times this trickster acts like a guide, shows us new maps, new perceptions of reality, that delight our senses and our intellect, and satisfy our need for learning how our universe is created. But often the trickster's maps fool our senses and our intellect. Indeed, speaking for myself, the trickster chides me. He lets me see that no matter how many times I cross the border using my tools and maps, I will eventually be fooled. My preconceived notions of reality, wherever I go and into whatever relationship I enter, defeat me or turn my new experiences into simply a repeat of past events.

You might think that changing my maps, redrawing my notions of reality, so to speak, and using different tools would bring me out of this predicament. If I am used to dealing with life as an intellectual affair, for example, if I think and use my senses to guide me through life, and I am fooled, I could suddenly choose to inquire into the nature of my life through my intuition rather than my intellect. I could still be fooled, however. Intuition can't always be trusted, nor can I always go solely by the feelings I have about a situation. If I could, a clever con man or woman would never fool me again. But our senses often fool us, and the con artist knows this too well. And so does the trickster within each of us.

Thus it is not easy to learn to use the new alchemical force of creativity, to become aware of the mind/matter split and aware when the unconscious mind influences our body. We must use all of our tools. And we need maps to guide us in the use of these tools. How we *use* the tools and maps is really the alchemical art. At times we must weave a delicate web, make a new map which crosses from one function to another. At other times we must let the universe itself tell us what comes next in the story of our life. Either way, the story will only proceed creatively when a hidden passageway is discovered. Without passageways life becomes terribly boring and static. We must be ready for the unexpected. We must keep an open mind and watch for it. Doing so whets our tools. This is the real meaning of "the philosopher's stone"—a means to sharpen our living tools of alchemy. Having whetted tools gives us the way to move through life in continual recognition and awe of its fundamental mystery.

Creation's Doorways

As we inquire about creation, Lenin's motto may come to mind. Perhaps we will make a step forward and grasp a new idea, and then later as we mull it over, we may find ourselves in confusion as we take our two steps backward and reconsider it. What is creation? What was uncreated before there was creation? Asking these questions may, surprisingly, bring us to their antithesis: What is *not* creation? This is a typical stumbling block that we encounter when we attempt to create something. When something new arises, its antithesis also arises. For example, think of any new idea you might have had in the past few days. As you pictured this idea, didn't you also bring forth the idea that the new idea wouldn't work?

Creation makes something appear, brings it out of the shadows, makes a distinction, shows something that wasn't previously showing, or enables us to see something new. Whether it's a star that appears for the first time or the birth of a child or a new idea or new concept that suddenly becomes part of your life, creative action surprises us.

Creation involves mysterious physical principles. We really don't know how anything comes into being. If we knew, if we could trace everything back to some fundamental law or idea, then, it, in a sense, wouldn't be creation. It would be simply a change from one thing to another, following some law or some course of action. For example, we might be surprised the first time we open up a refrigerator door in which we've placed a bottle of water to find that it now contains a rocklike, cold material called ice. Until we understand the laws of transformation that change the water into ice, we may think that the ice was created in the jar and the water somehow vanished. Before we understood the laws of transformation governing phase changes between liquids and solids, we would see this as magic, as creation. But our map, our notion of reality, based on our physical senses and physics theories, has provided us with an understanding that water freezes into ice.

And this is what science is—anti-creation. Scientists attempt to eliminate the idea of creation and understand everything in terms of transformation. They want to know how one thing changes into something else. They are not concerned with how one thing gets created from nothing, because there is no scientific way to grasp this. So, we might say the reason that there is a problem between science and mysticism is that the mystics are looking at creation from the viewpoint of awe and wonder and want to see everything as creation and nothing as transformation. The mystics might say that the laws of transformation are themselves merely illusions we create in our minds because we human beings want to have some control over our lives. We want to make sure that nothing gets created from anything. If something did, then the reverse is just as bound to occur. What got created could get un-created, it could be destroyed, vanish into nothing once again. And who wants that?

We want to have some semblance of inertia, something which holds us to the ground, and we want to know what that ground consists of. This desire makes anything come into being. The trickster is not a perverse being standing at the doorway to creativity. The trickster is our image of our self. Thus the trickster can keep the universe together by offering proper resistance, but can't completely block the creative impulse.

We are always interested in what's hidden or what a concept or an idea means. A clever magician knows this and is able to fool us because we are accustomed to seeing the world in a rather specific societal-bound manner. The magician captures our interest by implying that something hidden is about to be revealed.

Life is mysterious. It consists of a continual sequence of fundamentally unpredictable creative actions. No one understands why anything is the way it appears to be. No one understands why anything "is" at all. Creation itself, something coming from nothing, the arising of distinction, separability, all the ways we describe this process, are actually hidden from us. The magician is clever.

Face the door between the real and the imaginal and there the trickster stands. The trickster will lull us to sleep with the story of our life. The trickster will render us unconscious with the inert beliefs that we are what we think we are. What makes us unawake? It's simple. We are unawake when we are caught by the inertia of the consistency of our own reality—like trying to walk through a muddy forest. We walk around with the unconscious ideas that we've been taught, which define us as separate, distinct: "I am this; I am not that. I am good at this; I am not good at that. I am wonderful; I am terrible." These ideas have been ingrained in us since childhood and are reinforced by others, often family members, who have known us for a long time. But, they don't know us! They only know what they care to see and think about us. And we don't know them! We only know what we care to think and see about them.

These are the story lines of our existence. They are not mere wisps of non-inert energy. They are weighty. We embody them, so to speak, and after a long while we carry them with us as if we are walking around with a mass on our shoulders. We have each bought and sold ourselves on a certain story and in their discomfort they bring us a sense of comfort. Any idea or thought that runs counter to our story, even if it could improve our financial or emotional situation, we quickly push down into a safe comfortable zone of unthinkability. We have learned, without thinking, how to transform information into matter. This trick makes up the rules governing our unconscious mind.

In order to break free, in order to have a new experience, a shamanic awakening, a new vision, we have to break free of the illusion that we are separate from anything else—in particular, that we are separate from what we desire. This has been, is, and will always be, the secret of creativity: breaking free of separation. If we don't do that, we don't create. We must tap into the imaginal realm to create a new story line. We must return to unity in order to separate again.

hay

ה

Hay represents Life, an all-inclusive notion. When there is response (dallet) to motion (ghimel) of material (bayt) moved by spirit (aleph), we have Life. Thus the first four letters—aleph, bayt, ghimel, and dallet—are the seeds for the primary seed of Life.

Hay is spoken like a breath. The newborn child comes to life with a breath (hay) caused by its response (dallet) to the motion (ghimel) of the doctor's hand (bayt) on its backside, awakening in it the spirit (aleph).

Chapter 5

Life: The Body in Mind

Self-awareness, then, was simply a function of matter organized into life; a function that in higher manifestations turned upon the very matter that bore it and became an effort to explore and explain the phenomenon it displayed—a hopeful-hopeless effort of life to achieve self-knowledge. Nature turned in upon itself—a project doomed to ultimate failure, since Nature cannot be resolved in knowledge, nor can life, in the last analysis eavesdrop on itself.

Thomas Mann
The Magic Mountain

Warning: This chapter is an active parallel-worlds quantum-mechanical story. As such, many words and phrases have two or more meanings running in parallel lines. These different interpretations can appear to the mind separately or in various overlaps of meanings where no distinction into separate or "atomic" meanings occurs. So, while keeping all parallel lines open, maintain the tracks separately. Note when a shift or quantum leap in meaning occurs, when the mind realizes a distinction has occurred. Mind the gap or fall deeply into the imaginal realm of total acceptance.

But don't fear, for the trickster always stands guard.

And Now, To Our Chapter . . .

I am a body. At least that is how I scientifically think about my physical self—as an entity, a substantial being, a living physical object. But I also have feelings, thoughts, and sensations which tell me that I am not just my body.[1] And, I intuit that at some deep level, beyond what I can see or feel, I am not just my feelings, thoughts, and sensations either. My intuition tells me that I am more than my body; more, even, than my feelings, thoughts, and sensations. Yet if I believe the standard models of scientists, as discussed in earlier chapters, my experience, my intuition, is not valid. I am my body only. Nothing more. Such models constitute the materialist myth of science.

According to this modern materialistic scientific thinking, somehow I—the undefinable consciousness that provides me with awareness of my body—am formed from a complex of multiple, dumb, and simple processes. These involve photons of light, subatomic particles like electrons, atoms, molecules, cells, and organs, particularly the brain and nervous system. Finally at the end of this increasingly more complex and multiplying of processes, from out of the dumb and simple, emerges my whole intelligent and complex awareness of my body and my mind. There *I* am—emerging from a bubbling cauldron of electrically burning carbon-based slime. In less than three million years of evolution *I*, a descendant of cooked DNA soup, at last walk a land that, when I was soup, was under steaming water. My mind and my body are nothing more than some complex system of over-heated, telegraphing, microscopic hunks of electrified dead meat.

However, materialism and materialist-based science have radically changed over the years, and so has our own thinking about the mind-body relationship. In this chapter we will explore this radical transformation and apply some of the ideas concerning the new alchemy we have seen in the earlier chapters.

What Used To Be Isn't "What Is"

It used to be that we believed big things (like bodies) were simply constructed from other smaller things (like subatomic particles). A big thing behaved in a certain complex way because its smaller things behaved in simpler ways. Thus a human organ functioned the way it did because its molecules behaved the way they did. And molecules mindlessly behaved themselves because they were composed of equally mindless atoms—fundamental units of dead material substance inattentively obeying the mechanical laws of the universe as first pictured by the Greek philosopher Democritus and later by Sir Isaac Newton.

Ultimately all things in the whole dead universe behave mindlessly, following well-trodden paths, irrigation ditches, hoed in the space-time mechanical garden-universe that Newton so carefully cultivated hundreds of years ago.

Of course biological systems, on the other hand, are special. They are alive. Yet based on Democritus's ideas, we ultimately explain living systems as emerging from patterns made by dead mechanical pathways that their simpler molecules follow. Of course mapping these pathways would be a monumental task, impractical to say the least. But today, with our rapid advances in knowledge of DNA and molecular science, and with the amazing advances in computer technology, one might begin to think that such an undertaking is possible.[2] Even if it is, however, no one has yet seen an equation walk. Life and consciousness are still a mystery. And, as I said, materialism has had a rude shock delivered to its system: The simplest units of existence, subatomic particles and atoms made from them, do not follow the well-cultivated paths of Newton's garden. Instead, such objects—if they still can be called objects—seemingly follow multiple paths as they move from place to place.[3] Not only that, but somehow the mere act of observing such particles upsets these paths and alters their history.[4]

So today we are no longer faced with a universe of particles that exist objectively, independent of the observing instrument that attempts to carefully map their comings and goings. We now see the universe and all things in it very differently and, if I may add, a little strangely. This new vision comes from the discovery of the quantum nature of matter and energy, and it has changed the way we see the body and the mind. Einstein once pointed out that it was impossible to speak about matter without talking about space and time. I wish to indicate that not only is matter impossible to deal with without space and time, but it is also impossible for it to exist without mind. And mind—my mind and your mind—cannot exist solely within the confines of a body.

My views presented here, although based to some degree on my knowledge of physics, may appear, from a pure physics standpoint, as somewhat new, even bizarre. They are founded on a number of my own experiences dealing with the study of the relationship between consciousness and quantum physics. So, as I ask, "How does consciousness appears to arise in the body as an *I*?" my answer encompasses the following:

1. quantum physics—particularly the observer effect and the principle of complementarity;
2. a model of body and mind based on "story lines" or "scripts" that become enfolded in the cellular components of the body through the simple process of observation;
3. the location of the observer's mind or the search for the *I*;
4. mythic reality, and;
5. the nature of time.

This enfolding of mind, matter, and myth creates the *I* as a living, self-observing system in time. The *I* originates, however, beyond space, time, and matter/energy.

The Quantum Physics of Experience

Where, when, and how do consciousness and quantum physics overlap?[5] As is well known in quantum physics, the answer is specifically in the "observer effect," mentioned above.[6] According to quantum rules, an unobserved system such as an atom or subatomic particle does not exist as a "real" particle. Instead it exists as a ghost-like cloud of possible physical particles. In the jargon of quantum physics, these possible physical particles are called "states" and, in general, they are the yet-to-be observable and thereby measurable attributes of our experience of any physical system. As such they reflect tendencies toward existence rather than physical existence itself. These tendencies or probabilities, therefore, connect psychology to quantum physics.

A probability is "a tendency for something," as Heisenberg described it:

> It was a quantitative version of the old concept of "potentia" in Aristotelian philosophy. It introduced something standing in the middle between the idea of an event and the actual event, a strange kind of physical reality just in the middle between possibility and reality.[7]

For example, there are tendencies or states corresponding to the locations of an atom in the body or anywhere in space. Before observation, these states spread like an expanding cloud over space and time. To imagine this, we would need to picture them as billions upon billions of tiny dots filling out space and moving through history. Each dot represents the tendency of the atom to manifest at one specific point in space and time. Upon perception, observation, recognition, cognition, or registration of some physical recording device, suddenly this cloud of ghostly dot possibilities evaporates, leaving the atom by itself in a single physical location at a specific time. Thus a tendency becomes an actuality. In this picture, the billions of dots would all vanish, leaving only one.

When this sudden evaporation occurs, the atom physically appears and, most importantly, the observer of that atom measures it. This cognition occurs at any time and at any place, even outside of the body. Of course, this leaves open the question: Where, oh, where can the observer's mind be? Or for that matter, what constitutes an observer's mind or a cognitive experience?

Some physicists believe that the observer's mind is not responsible for the sudden evaporation of the probability cloud of a single atom. In fact, no one has really ever seen an atom. What physicists see when they "look" at an atom is a record made by a measuring instrument—the track of the atom, so to speak. Instead of the observer causing the cloud to evaporate by noticing where it is, these physicists believe that somehow it is this measuring instrument that causes the field of dots to vanish, save one. And many physicists take this as an indication that the action of "dot evaporation" lies outside of the realm of physics, that all an observer does is confirm what an instrument tells him or her. Other physicists believe, however, that the sudden evaporation really only occurs when the observer consciously notes the measuring instrument's indicator. They would say it doesn't matter where or when the observer comes in, the act of observation by a knowing, conscious being ultimately causes the cloud to pop into a single drop—even if a large measuring instrument interposed between the object and the ultimate observer pops the cloud.

The "pop" is a thorny knot in the side of physics. No one knows just how this sudden popping from the imaginal possible to the real takes place. There is nothing in quantum physics itself that predicts this occurrence. Yet, this sudden "pop of reality" is the basis of Werner Heisenberg's uncertainty principle, Niels Bohr's principle of complementarity, and the consternation of physicists worldwide.

Also called the principle of indeterminism, the uncertainty principle reflects the inability to predict the future based on the past or based on the present. Known as the cornerstone of quantum physics, it provides an understanding of why the world appears to be made of events that cannot be connected in terms of cause and effect. In essence it says, "Let there be lots of dots."

The complementarity principle says that the physical universe can never be known independently of the observer's choices of what to observe. These choices fall into two distinct or complementary categories of observation. Observation and determination carried out using one category always precludes the possibility of simultaneously observing and determining the complementary category. For example, the position of an object and the path it follows through space and time (the object's momentum) are observations in complementary categories and so cannot be determined simultaneously.

Consider figure 5.1. Although it is only a visual analogue to the complementarity of a real quantum process, it may indeed reflect quantum physical processes in your brain. For the moment cover up the right figure of the drawing and only look at the left figure. Look at the thick white lines. Do you see a cube? Which face of the cube is in front? Are you looking down on the cube or looking up at it? Does it suddenly jump? By looking at the drawing in this way you see a jumping cube. But hold on. Can you also see that the left drawing consists of a patchwork of geometric black plane figures painted on a white background? If you can't, cover the left figure and look at the right figure of the drawing. This illustrates a complementary way of viewing the picture on the left. How you see the left illustration depends on how you choose to see it. And you can't see it both ways at the same time. You either see a cube or you see flat shapes against a white background.

Now think about this. Where did the cube jump? Where did the geometric figure of the cube emerge? "Out there" on the page or "in here" in your mind? If the latter, then where is your mind located? (We will return to these perplexing questions later.)

In fact complementarity affects how we all see, think about, and feel about the world and perhaps more importantly how we see, think about, and feel about ourselves. How do you imagine yourself? Are you primarily a thinking person or a feeling person? When asked a question do you respond, "I think that . . ." or do you respond, "I feel that . . ."? Jung pointed out that human personality has either the thinking or feeling function well developed, but not both. Thinking and feeling are complementary to each other.

(By the way, if you haven't seen the cube jump yet, you are probably, as Jung would have put it, a thinking-type individual. Don't worry if it hasn't jumped as it does for many supposedly feeling-type people. If you keep looking at it, it will eventually jump.)

Figure 5.1. The Paradoxical Cube. In the left figure, whether you see a cube or black pieces against a white background depends on you. This is the principle of complementarity in action. In the right figure, black pieces against a white background is more readily apparent.

In a similar manner, our choices alter the physical body. How do you think about your body? How do you feel about it? Can you separate your thoughts about your body from your feelings? How do they differ? How are they the same?

Getting back to the pop, look at the drawing again. That sudden jump is a visual metaphor for the quantum leap or the pop. The pop has given rise to many different interpretations besides the two mentioned above (the observer did it or the instrument did it); all, with the exception of one, require metaphysical belief systems which lie outside of the laws of quantum physics. The one exception is perhaps the least acceptable, although it is the only one that remains within the bounds of quantum physics: It says that the pop does not occur. And this interpretation may explain how the mind and body interact and become coupled together.

Many Worlds Braided Rather than One World Popped

The "many worlds" or "parallel universes" version of quantum physics states that the observer, in observing, is actually becoming a part of the observed by noticing and remembering what he or she experiences.[8] If a quantum system is capable of being observed in one of several possible states, then when an observation occurs, the system enters all of these states and the observer's mind splits into a companion state associated with each possible physical state of the system.

Look at figure 5.1 again. Try once more to see the left figure as a cube. When you see that you are looking down on the cube, you and it have entered one parallel world. When you see that you are looking up at the cube, you and it have entered another parallel world. In each world, the cube exists with its position determined as you see it, and you also exist with a memory caught by your own observation. Your memories/observations of the cube's positions and the cube's positions in the two parallel worlds form the many-worlds basis of the mind-matter interaction. In each world, your memory is paired with one of the cube's possible positions; and you and your memory exist in each world unknowing of the other's existence. In each world one possibility emerges as real and the other as imaginal. Hence we have an explanation of mind (the imaginal) and matter (the real) emerging at one and the same time.

Paradoxically, if there were only a single world, this communication would not be any problem for the observer. After the observation, the mind of the observer is no longer free; it has been split by the system it observed. Thus the observer is caught by the observed and paired with it, much as two people become paired when they fall in love. It is not that the other possibilities of the ghost cloud suddenly vanish while one of the possibilities materializes; it is that all of the possibilities are observed in parallel worlds and the observer's mind is in each of those worlds.

This is schizophrenia on a madcap scale indeed. Thus in each probable world where a physical object exists in a specific state, an observer observes that object in that state. This interpretation says that all possibilities exist simultaneously and all possible memories of the object also exist simultaneously. In this way mind and matter are intertwined in an infinity of braids, all arising from a single encounter of mind and matter.

Imagine all of the events of your lifetime laid out on a map. Not only would all of these mind-matter links composing your life exist now—which would be seen as a spread of points over a map at a single time—they would also be spread out over the map of time, frozen in space-time like bugs in aspic. And that little statement says a lot. These intertwined mind-matter states exist now, they existed before, and they will exist after.

These intertwinings can be pictured as a field of worms frozen in the space-time ground of being. Since they are spread out in time as well as space, they are actually multiple histories of the relationship of the observer to the observed. They exist as an overlap of story lines extending back to the beginning of time and forward to the end of time. These story lines can be imagined as twisted braids of mind and matter woven throughout the history of sentient intelligence.

Again, look at the left side of figure 5.1. See the cube. As it jumps, you are experiencing the twists and turns of one small part of your own story line. In this case, the story is simple. It only has two plots: You are looking up at the cube or you are looking down at it. But, of course, there are lots of stories. And these stories form epics, which ultimately make up the various species of life on the planet. Such story lines also make up the human body and, for that matter, all things in the universe. In this picture, all things are alive since they contain memories embedded in matter. This includes rocks and all material elements. It includes the earth and even its molten core, not to mention biological systems such as plants and all other animals.

And, as I pointed out, of course these story lines exist in the human body. The twists and turns of the DNA molecule are perhaps alchemical physical manifestations of the twists and turns of the mind-matter story lines. When these story lines occur in a human body, I shall call them "scripts."

A script is a braid of possible past stories. It contains the history of the human species as well as the record of what you ate and didn't eat last Tuesday. In each story an observer memory and an observed event relate. The scripts follow all possible paths in time. Thus every bit of matter in the body has its story to tell, for it has its own ever-present storyteller—the living, conscious body. This body has mind intertwined all over and through it. The whole body acts as a great recording device. It has made records of all possible pasts that could lead to its present, and it anticipates all possible futures. The futures exist in the present because of the probabilistic nature of these story lines. They are, after all, scripts of possible pasts and for the same reason visions of possible futures.

We might call these scripts of the body the "dreambody," a term originally suggested by Arnold Mindell. In his book, *Dreambody,* Mindell points out that the various traditional disciplines each have their own view of the body.[9] In other words, the body is seen as one thing by Buddhists, another by Christians; one thing when seen by Western practitioners of medicine, and still another when seen by physicists.

So, we might ask, which is the real body? Surely it must be the body we see in the mirror. Not so: that "corpus reflects-us" is still just another observation. And, as we have seen by looking at the cube drawing, it is what the observer chooses to look at that determines what he or she sees.

In his dream and body work, Mindell, who describes himself as a process-oriented psychologist, found that dream images and symptoms of illnesses point to tendencies for particular psychophysical processes. Dream images and symptoms create an outline for stories telling you about yourself. For example, are you feeling some discomfort in your body at the moment? What comes to your mind when you feel this discomfort? Use your imagination. Perhaps you have a stomachache. Perhaps this ache brings up a memory of your childhood when you had a similar stomachache. These tendencies may manifest as body problems, relationship difficulties, dreams, and/or synchronicities. They reflect a person's relationship with events in space and time. However, since these are *tendencies* or *probabilities* for events, they only *suggest* changes rather than command them. Mindell suggests that these tendencies or probabilities connect psychology to quantum physics.

Here I wish to suggest that the images we have of our bodies, the pictures that emerge, come from our scripts—the stories we have come to believe about ourselves. These scripts are rewritten anew or refreshed every time we move, every time we think about ourselves, and even every time we fantasize about ourselves. The stories are embedded in our bodies in much the same way that audio or visual information is embedded on a movie or videotape. New experiences are recorded everyday and matched with old ones. When we enter into a new experience we not only perceive what is "out there," we also project to the outside world what is "in here" and we make a comparison of the scripts. Does this new story map well with my old one? Should I run away, or should I stay and play?

In my model, our scripts are matched up with our sensations of the outside world. Remember, scripts are the intertwinings of observer states with physical states of cellular matter in parallel worlds. As such, they not only contain memories of what was, but also of what may have been. And, they also contain stories of *anticipation* of what will be. Given this, how can one's "real" experience of the "out there" world arise? How are we able to experience anything, whether it is information coming into our nervous systems and brains from the outside world or data that is generated in the other cells of the body? What I am really asking is this: How does consciousness arise? Where does it arise? Seemingly it occurs in the body, most likely in the brain.

The perception of bodily reality as it supposedly occurs in our brains and nervous systems can be, I believe, described as attention paid to these parallel world scripts and projected from all of the cells of the body into space and time. We need the parallel worlds in order to determine which possible story is true and "really out there." Without the parallel world possibilities, we would not be able to change the ways we experience the world because we would only have one fixed memory with which to compare things.

For example, Nobel Prize-winning physiologist George von Békésy found that subjects, stimulated by physical devices but deprived of their visual sense, actually felt sensations in a space where no parts of their bodies were even present.[10] Von Békésy placed vibrators on the knees of subjects. He then asked them to part their knees. As the vibrational frequency was altered, the sensation appeared to jump from one knee to the other and then, at certain frequencies, it appeared in the space between the knees. The vibrations produced interference patterns, that is, competitive scripts in the brain of the observer, and thus recreated an experience of objective reality "out there." In addition, Benjamin Libet of the University of California, San Francisco Medical School, in a series of remarkable experiments, established that sensory information is also projected from the brain backwards through time.[11] To be fair to Libet, he didn't actually state this. He did say, however, that sensory information appears in the consciousness of the observer "referred backwards in time." By this he meant that although the subject's brain didn't actually indicate that awareness of that stimulus occurred until some period of time after the stimulus (roughly a half-second), the subject experienced the stimulus as if it happened very nearly at the time the stimulus was applied (roughly about ten milliseconds later).

Libet believes that space-time projection is essential to the human mind-body connection. As he put it in a conversation I had with him some years ago, "Referral in space is, in principle, similar to referral in time. We discovered referral in time. There is referral all over the place." Thus the sensation of feeling something "out there" in space or backwards in time is then comparable to the sensation of seeing something "out there" in normal vision or feeling that someone is out there when you can't see a thing. In fact, in physics today we believe that space and time may not be primary concepts. This indicates that, in some way, we project space and time as well as the spatial and temporal extent of objects. Perhaps we do both at the same "time."

So, where is the objective world actually experienced? When? In our brains, in our bodies, or out there and back then? Even if we answer these questions, we have an obvious problem: Where is the observer? Indeed, where experience takes place and where the observer exists are the most difficult things to talk about. Where is the homunculus? Where is the "person" who experiences the outside world? In the brain? In the body? In the whole universe? In all of my research, I have yet to find the location of the "observer" of reality in the brain or the nervous system. I haven't found him or her in the body either.

Just as, according to quantum rules, the "true" object seems to fade away—like the cat's face in *Alice in Wonderland*—I conclude that there is no one there to observe it. There is no person in the body or in the nervous system. Referring to what we would call the ego, the Buddha taught there is no *I*. He also saw that there are no objects out there independent of this *I* who does not exist.

Not much to stand on is there? The Beatles were right: "Nothing is real." The French perhaps got even closer to the truth when they coined the word *personne* to signify *nobody.* But if there is no witness, no fundamental observer, then what is going on? Don't get the Buddha wrong, there is something going on, but it is not as it seems to you, for "you" don't really exist.

But if you don't exist, what does? Well, it seems that the scripts are real, written down in an imaginal realm that may be more real than what we perceive.

The Imaginal Realm

In the scripts model, the observer, by observing, actually becomes the body. Bodies are, in a sense, "living scripts." At the level of the body, the observed and the observer are the same thing.

Now, as I said, there are an infinite number of parallel scripts. This idea comes from the parallel worlds theory in quantum physics. But, can we experience parallel worlds simultaneously, or do we have to split off into myriads of parallel realities? The answer is yes to the former, no to the latter. We can experience simultaneous parallel worlds without splitting off. When such an experience takes place, we have entered a form of timeless mythic reality. No pop actually occurs when we see parallel worlds simultaneously, as, for example, when we see the above drawing (figure 5.1) as a patchwork of geometric figures. So, nothing changes; everything just is. But, when we see a cube pop into existence, we see either one world or the other because our partner-self also sees either the opposite world or the other. The pop that appears to flip you into one world or the other is the quantum leap, which signals the passing of time and the splitting of the unified mind at large into a single mind perceiving one world.

The overlap of parallel worlds (seen in figure 5.1 as a patchwork of geometric figures) can be imagined to be the mind without subject-object distinction. It is the realm of pure subjectivity; the realm of consciousness without an object of consciousness. Here there are no jumps or pops. We have the world of all possibilities arising simultaneously and yet nothing is realized. For there is no realizer and there is no realized. This overlap of unobserved possibilities is as ontologically "present" as the geometrical patches are present in the drawing. It can be thought of as real alternatives in an imaginal realm that persists beyond time. In our new alchemy it can be imagined as the field of aleph containing the possibilities of bayt brought into time and space through the action of ghimel and recording into memories as the field of dallet. As such the life field of hay is realized and made aware.

The overlap of parallel story lines or body scripts resembles the Australian Aboriginal concept of "dreamtime"[12] and Henri Corbin's concept of the "imaginal realm."[13]

As new as the dreamtime concept of this reality may appear to us, Australian Aborigines claim to have "memory" of this realm dating back nearly 150,000 years. From this realm, a long time ago, the world of mind, matter, and energy arose as a dream of the "Great Spirit." Thus aboriginal thinking suggests that the universe or God is itself dreaming into existence all of what we experience.

Figure 5.2. One You, Many Possibilities. Looking at this piece of Aboriginal artwork, I can imagine that the two snakes indicate the braiding together of stories, one starting out in the past and the other in the future.

Henri Corbin, noted scholar of Islam, was the first European author to coin the term "imaginal realm." In his view, this realm is ontologically real and as my own research into the nature of shamanism suggests, it just might be more real than the reality we perceive.[14] However, this reality usually exists beyond our normal waking perception as a real and yet timeless realm. By that I mean that the quality of time passing vanishes, things do not change, everything shimmers as present.

vav

ו

Vav, or waw, represents the copulative element, the "male" energy, the primary or fundamental act of fertilization. Male in both its action and its form, vav performs by copulative reproduction, producing seeds.

Vav is an archetype for potentiality. Vav appears in multiplication. It means and *in Hebrew.*

CHAPTER 6

Endless Fertility: Is the Force with Us?

> *What was life? No one knew. It was undoubtedly aware of itself, as soon as it was life; but it did not know what it was.*
>
> Thomas Mann
> *The Magic Mountain*

What unknown force pushes living molecules into their evolutionary future? Does this force connect us to God? Does it enable us to make invisible links to the space-time continuum? Is that why quantum uncertainty exists—because God uses human consciousness to co-create the universe? Or does God make us the unwilling embodiment of God's spirit, knowing that we cannot manifest anything on our own but merely act as robots under the watchful eye and hand of our maker? Think about it. Does a force of destiny, an invisible watchmaker's incessant winding, keep us on our toes? If so, we might believe that we have no free will, no independent minds, and no abilities to make choices.

But, is this true? Are we no more than robots following a subtle force that engulfs us so completely that, like fish in the ocean not knowing of the existence of water, we are unable to discern it? Do we have any choices in this evolutionary game? Perhaps. If we make the seemingly "right" choices, we sense no resistance. We go along blindly with the incessant flow of time.

But, if we make the "wrong" choices, we ultimately suffer. We end up being pushed by the force like a current in the ocean moves a fish or a breeze in the troposphere nudges a high-flying bird.

How can we uncover such an alchemical force—a *time-wind* pushing living molecules into their evolutionary future? In a series of reports, visionary physicist John A. Wheeler may have told us how. He reviews what quantum physics and information theory have in common regarding the age-old question, "How come existence?" He concludes that the world cannot be a giant machine ruled by pre-established law. The laws of science as we know them provide mere idealizations concealing the true information source from which they derive. Fundamentally there is no universe of physical matter-energy without the *elementary act of observer participancy* that Wheeler epitomizes by the phrase, *it from bit*—or in Biblical terms, how a thing arises from a word. In other words, each of us must take responsibility for making the world the way it seems. And the way it seems springs from our actions, which are, in turn, conditioned by words and symbols.

Even so, why do we choose to do the things we do? Does God have anything to do or say about the actions we take? It seems apparent that some, perhaps gentle, force is guiding our hands and minds. How does this force come into being? How can we discover it? Somehow, we need to understand not only the players in the game of life—molecules of life following complex but necessary patterns—but also the *rules* themselves running those patterns. How do these rules arise?

Perhaps these rules are somehow contained in coded messages distributed throughout space and spontaneously arising at specific instances in time. Perhaps as the universe expands, these rules appear to come into being at the "edge" of space "where no one has gone before." In this chapter we will explore this possibility, how such rules could arise, where they come from, and why they must be present in addition to the seemingly random but deterministic causes dictated by natural selection. One could say this is a chapter about the *quantum rules* of the alchemical game of life.

According to Hoyle

In his book *The Intelligent Universe*, Sir Fred Hoyle argues that although we are a species capable of wondering about life, the universe, and everything in it, we hardly ever take stock of our purpose in this great chain of being.[1] We raise our children, make our living, and seek "better" lives for ourselves, questioning little, paying little attention to our neighbors. Occasionally, however, we do look beyond our individual lives and watch out for that fellow across the street, across the nation, or even across the world. Why? The sciences of biology and physics lead us to believe that this has nothing to do with morality or compassion, that it has to do, rather, with the pure unadulterated biological need to reproduce ourselves, to insure the existence of the next generation. But what purpose would that serve? This endless chain of reproduction would merely go on and on into the infinite future. Does that really mean anything?

According to Hoyle, science has no other explanation; continued production is the only answer. The machine must go on mindlessly and amorally churning out the next product whether that machine is based on the nuts and bolts of sexual urges acting as the whips of continued successive generations, or the rack and pinions of Newtonian mechanics acting as the grinding reticulated backbone of industry.

If that's all there is to this then, Hoyle asks, why do we have any sense of morality at all? Biologists argue that morality is simply a nonscientist's way of understanding the biological fact that animals act in tribes, packs, schools, or hives to insure survival—they survive because they act and work together. The Communist party and dialectical materialism pops to mind here. Hence, we are concerned for our fellow human beings because that is the way we ultimately insure our own survival. If we could survive without him or her, we would not act cooperatively and morally.

But surely, if mere species survival was all there was to it, there must be other biological factors we could have inherited to increase our chances. As we evolved along the long line of time, dating back three million years or so to our earliest ancestors, why didn't we develop the ability to run faster or to fly as birds, as hares and eagles do? Surely these characteristics would have assisted us even more than mutual cooperation. But we don't have these characteristics. True, we might desire them and turn that desire into machines that artificially give us great speed and the ability to lift ourselves off the ground—which, of course, we have done. But our desire, so far, has not provided us with these physical characteristics. If natural selection was at work during our long history of evolution why wasn't the possibility of wings on our backs or long rapidly moving legs encouraged? Why didn't the "blind watchmaker" choose these characteristics for humans?

Today people believe that Darwin's theory of biological evolution through natural selection insures and explains the survival or extinction of any species and that natural selection insures its evolution or destruction. The theory of natural selection posits that creatures within a single species come into being with a random variety of characteristics, most of them good and some of them probably neutral or bad. Of course "good" and "bad" refer to the ability of the creature to adapt to its environment. Creatures with characteristics appropriate to their environment would survive, while those without characteristics appropriate to their environment would become extinct. Thus arises the notion "survival of the fittest." But, is this merely a rationalization of human behavior?

Thirty years ahead of Darwin, in 1830, scientists already knew the ideas contained in the Darwinian theory. So why weren't the ideas popular then? Because society didn't yet need them. Darwin's theory became useful, and thus popular, as the industrial age emerged. Companies were competing with fierce abandon, nations in Europe and elsewhere were seeking *lebensraum* (literally, "living room"), and kings were jockeying for advantages by playing warfare games with people, not pawns, on the real chessboards of life—their own country's inhabited landscapes and cities. In other words, Western society was rampantly exploiting the world and had begun what we now call the dog-eat-dog world view of commerce. And Darwin's *On the Origin of the Species* was just the right bible for this new enterprise. You see, science and industry go together hand in glove, not only in devising the technical means of production, but also in the employment of the alchemical forces of the imaginal realm that fuel such enterprises. Thus, particular heads of industrial states of a society, reading Darwin, create the industrial environment as "proof" that the theories are correct. These heads then adopt the Darwinian theories of science and make them popular with people ready to except the theories and advocate them, further fueling these machines of society. This circuit becomes dominant and stable wherein what we think is supported by the environment we live in, which in turn limits and directs our thinking, thus enabling species survival.

We see an example of this everyday. Companies believe that to survive they must improve their products. Within months of its introduction to the buying public, nearly any product we see on the store shelf will now bear the words "new and improved" on its label. Changing a commercial product for the better is considered a necessity for business survival, a notion that fits with Darwinian improvement of a species through natural selection. Just as you or I select the "better" product off the shelves thus insuring its continued production, Nature selects which species shall go the way of the dinosaur and which shall go the way of the housefly.

In other words, Darwin's theory became popular because it rationalized the human need to compete, which in turn rationalized human greed and fueled the industrial revolution.

Blind Lady Luck and Molecular Darwinism

Having its roots in the pre-industrial European environment, as explained above, today Darwin's theory seems to be present even at the molecular level. Go back a few hundred years before that time and you'll see a tendril. Around the time when Anton Van Leeuwenhoek invented the microscope, in 1673, people were led to believe that species were formed whole and separate from each other, more or less as described in the Bible. Human beings, for example, were formed with distinct characteristics, which led, naturally, to the ideas of race, superiority, and a hierarchical society with divine rulers at the top and not-so-divine "untouchable" laborers at the very bottom. This layer model insured that society had a divine right to its ruling classes and all of the layers in between, that each person, no matter which layer he or she happened to be on, should be happy. For after all, to be at a certain level of the layer cake meant one was right where one should be. Thus, people are what they are and should be content to remain at the level of society where they find themselves.

Populist revolutions against this divine-order theory continually arose, particularly in France during the eighteenth century. With ease, royalty squelched these attempted revolutions of the divine order. But, Jean Baptiste de Lamarck, who lived through the French revolution of 1812, had a theory, most probably as a result of the spirit of the times, that ran counter to the "divine layer cake." Simply put, he believed that parents were able to pass on their acquired characteristics to their offspring. So, for example, because its environment did not provide the necessary ground vegetation, the giraffe's short-necked ancestor foraged for leaves in high-branched trees. It stretched its neck to get them. Its offspring were then born with slightly longer necks, and this cycle continued through time to the present long-necked animal. Accordingly, a tall coal miner who spends his time stooping in a coal mine should pass on his work-acquired stooped shoulders to his children. Eventually offspring from him should appear with shorter statures and bent over shoulders. But, of course, this is not the case.

DNA is the primary element of genes and thus genes are responsible for the characteristics of any species. We know that if we change the genetic structure of an animal we indeed change the bodily characteristics. But so far, there is no evidence that an environmentally induced change in a body, such as the loss of a limb, or a thicker tennis player's arm, or a chess player's developed brain, alters the genetic code. Hence Lamarckism, although steps away from "divine layers," appears nevertheless wrong—which is another reason why Darwinism *appears* to be a feasible theory.

"The one thing that makes evolution such a neat theory," wrote zoologist Richard Dawkins, "is that it explains how organized complexity can arise out of primeval simplicity."[2] To prove this, Dawkins and many other Darwinian followers hype the early experiments of Miller and Urey. In 1952–53 Stanley Miller and Harold Urey passed a 60,000 volt electrically generated spark through a bottled mixture of water, nitrogen, methane, ammonia, carbon monoxide and dioxide gases. They repeated this event over and over again for a period of several days. Their aim was to emulate the imagined electrical storm-driven atmosphere of earth about 4.5 billion years ago. After their experiment, they found that their bottle now contained sludge, and the sludge contained a number of different organic molecules—the kinds associated with living things. Among them were amino acids—the building blocks of DNA.

This experiment has been repeated a number of times and leads us to wonder: Since amino acids are the building blocks of proteins, and proteins are the building blocks of living things, perhaps this was how life itself began. Many tout this as proof of Darwin's theory of molecular evolution.[3] Accordingly, life would have evolved from simple dead molecules into the complex and living protein-based structures we now know so well. There is no evidence at all, however, that anything like a protein has ever been produced in the artificially created sludgy primeval soup believed to have existed at the dawn of life.

Indeed, proteins are themselves generated by the DNA of a cell in a complex series of processes. Some of these proteins are called *enzymes.* Today we know the molecular structures of perhaps 2,000 enzymes. They are metaphorically thought of as protein "weapons" used by a cell in its "battle for survival" against the environment. Enzymes provide way stations where other molecules enter, bind for a while, exchange energy, and then depart leaving the enzyme exactly as it was. Thus enzymes allow vital chemical reactions to occur in the cell, which enable it to continue existing, fighting against the slings and arrows of molecular warfare continually ensuing in the body. Without enzymes these necessary chemical reactions would not occur or would occur at such a slow rate that cellular life would be impossible. The food we eat, for example, would be useless to us without enzymes, for no reactions releasing energy from that food would ever occur.

But, I repeat, there is no evidence that any process, devised in a manner similar to Miller and Urey's experiment, created a single enzyme. So how, then, could enzymes have evolved naturally from amino acids? The organic sludgy soup theory of the origin of life is clearly faulty. Yet modern biology persists in this Darwinian fantasy, often claiming that mere chance alone would be enough to produce the nearly 200,000 amino acid chains required for molecules such as DNA and enzymes, to mention just a few. The argument goes like this:

Given 4.5 billion years instead of a few days and an ocean instead of a lab bottle, there would be enough time and space to insure that this would happen. But this is patent nonsense, as any good mathematical analysis would tell us.[4]

One could argue that perhaps our life-world is just one of a very large number of equally likely life-worlds and that even though the probability for our particular life-world is extremely low, the total probability for all equally likely life-worlds is high. In other words, the sludge could have produced different forms of life with different sequences that would have evolved into intelligent life forms *like ours*, but nevertheless quite different from ours. This theme is often shown on science-fiction television shows. And maybe it's true. But there is still a problem. Why Darwinian evolution? Why not unintelligent life forms merely mechanically adapting to changing environments much as a thermostat adapts to changes in the temperature in a room? Perhaps some computer-generated Darwinian models attempt to reduce evolution theory to just such a scenario.

So where does the new information providing the choices for natural selection come from? And how does this information come into the world? I'll tell you: New information is not generated in the past; in terms of space and time it must come from the future. Before you dismiss this theory, consider the evidence. Consider also why you may be feeling skeptical.

Information from the Future

We all seem to know what we mean when we say that an event has occurred. We mean that something happened. We also mean, although we don't usually say so, that someone has observed the event. We take it for granted that events *could* have occurred, even though no one was there to observe them.

After all, that proverbial tree we come across, lying on its side in the forest, *must* have fallen with a great crash of sound as it thumped the ground in its grand descent. Events such as sound-wave production must happen regardless of the presence or absence of beings with ears or other organs that can sense these events. Right?

Such a viewpoint is quite natural and is called *objectivism.* This philosophy states that all reality arises objectively, externally, and independently of the mind. Knowledge of this reality comes from reliably based observed objects and events—things that happen to these objects. But suppose science proved that objectivism was wrong. Then what? It would mean that all reality is not objective; that reality has to have a subjective quality to it (called mind); and that mind has to affect and possibly even change what we sense as objectively "out there." In brief, it would mean that there is no absolute "out there" unless there is a mind "in here" that perceives it.

Hence a perception of an event may act precipitously, actually bringing on the event and giving to the event its various observable qualities, including the biological characteristics seen in life forms. Though perhaps conceptually strange, such a view has been realized by the discoveries of quantum physics. These discoveries show precisely how events themselves, things that happen in space and time, are affected by the actions taken by observers. I'll have more to say about the timing of these actions in the next chapter, but let me indicate here one such outcome called the *Zeno effect.*

Suppose two elementary identical systems (like atoms or molecules) under our control are allowed to evolve in time in two different ways: one system is continually monitored as time passes; the other is only observed at the start and at the end of its evolution. Finally, the two identically constructed systems are compared. What we find is that they have ended up quite differently simply because of the number and frequency of observations.[5] The Zeno effect shows that simply by his or her choice of what, when, and how often observation takes place, the observer determines an actual history of the evolving object. For example, an atom, ripe to radiate energy, can be frozen by simply observing it repeatedly so that it never gets to shine its energy—something it would normally do within a few microseconds when only observed at the beginning and at the end of its radiating life.

Surely I must be talking only about atomic and subatomic events—those that occur on such a small scale of time and space that they would play no practical role in the evolution of a living species. Right? Actually, that argument, itself, makes the point: Since our human evolution is determined entirely by our molecular codes, DNA and the like, observations occurring on such a small scale as the molecular environment are destined to effect the evolution of a whole species. Indeed, a whole universe and all life within it could and would be affected by such micro- or nanosecond observations. In fact, practical applications of this now occur daily.[6]

Let's suppose that's all there is to it. On this scale certain kinds of observation occur and they encode the rules and instructions required by an evolving life form. This encoding enables the life form to survive and to reproduce. Thus an observer, by knowing just when, what, and how often to observe on this scale, could nudge a life form into a new action—indeed, with enough time, into a whole new species. All we would need for this would be an observer with novel information who had the ability to make it available in the environment. Then all the life form would need would be an already encoded instruction that enabled it to "listen" to that information. Since the life form could be anywhere at any time, the observer's mind would necessarily have to be everywhere in the environment.

I'll get back to this all-seeing observer in a moment. For now, let's look at how the life form "learns" this instruction, namely the instruction that tells it how to instruct itself by learning from the environment.

The first absolutely necessary code would be a statement that in essence says to the creature, "Absorb information from the environment." Can the theory of natural selection explain how such a code could have first been absorbed from the environment by a life form that did not have this coded instruction? Well, let's suppose there were ninety-nine nearly identical creatures with the characteristic don't-absorb-information-from-the environment (DAIFE) and one of them with the characteristic absorb-information-from-the-environment (AIFE). If the environment has no information or if the information it contains, if absorbed, causes this creature to die, the ninety-nine DAIFE creatures continue to survive and the one AIFE creature would probably die out. It would not survive since this particular characteristic, AIFE, either would not be needed for survival (assuming that AIFE replaced another characteristic needed for survival) or the information, if absorbed, would kill it. But if good information is there for the picking—for example, the ability to change with an environmental drop in temperature—the AIFE creature would be naturally selected by the simple fact that it was present when the environment was capable of feeding that characteristic. Consequently the ninety-nine DAIFE creatures would be at a disadvantage since the new information enabled future generations of the one AIFE to deal with temperature changes while the offspring of the ninety-nine DAIFE creatures could not.[7]

But why would nature randomly produce, even just once, the right characteristic needed for future survival? Of course one could guess that nature produces all of the possible characteristics it can. But somehow I feel this is incorrect, simply because there are so many different kinds of characteristics. The slightest numerical adjustment in any of them would render them useless. Yet here we are, evolving and hopefully surviving.

One answer to this question states that the necessary information comes from the environment—not the environment laid down by the *history* of the species, the planet, or even the universe, but by the *future* of all of these. In other words, *information flows to the present from the future.* Certainly we all believe in the value of hindsight. We believe that we can relive past events in our memories and see just how to change them to make them better or worse. Well, imagine that you are able not only to *look* back in time, in your memory, but to actually *go* back in time and make the adjustments you wished. See yourself nudging some event that had, in the past, dire consequences for you or for another. This nudge needn't take much energy; it could be a mere look on your face witnessed by another, or perhaps a slight change in your tone of voice when speaking to that other. Quantum physics allows us this type of mind-freedom, but it is not as simple as I make it. In other words, it is possible, in a certain quantum-physics sense, to alter the past by changing the way it exists in memory. It isn't that we really go back in time; it is more that we add details to the events that were not sufficiently specified at the time they occurred. For example, we could determine by which means an atom journeys from one place to another by altering how it is perceived at the end of its trip. One choice of that perception would "create" the past history quite differently from a second choice we might have made. In this subtle manner, we create a past from possible counter-factual histories.

If this is true, what is time really all about?

zayn

ז

Zayn represents energy in the process of breakdown, a state which enables new possibilities to arise. Zayn is pictured as all the possibilities resulting from the disseminating action of vav. Whereas vav is the particle-like production of seeds of possibility produced from hay (life), zayn is the wave-like field of possibilities grown from these seeds and may also be thought of as the primary or the main principle of indeterminacy.

Zayn sings "anything goes," and its tune may be truly zany.

Chapter 7

My Time Is Your Time: Anything is Possible

"Have you also learned the secret of the river—that there is no such thing as time?"

"Yes Siddhartha, is this what you mean? The river is everywhere at the same time. At the source and at the mouth. At the waterfall, at the ferry, at the current in motion, and in the mountains. Everywhere. The present only exists for it, not the shadow of the past nor the shadow of the future."

"That is it," said Siddhartha. "And then I learned that as I reviewed my life, it is also a river. Siddhartha the boy, Siddhartha the mature man, and Siddhartha the old man were only separated by shadows not by reality.

"Siddhartha's previous lives were also not in the past and his death and his return to Brahma are not in the future. Nothing was, nothing will be, everything has reality and presence."

Hermann Hesse
Siddhartha

According to Einstein's theory of general relativity, matter cannot exist independent of space and time. If any one of the three—matter, space, or time—is absent, they all are. Space is necessary in order for matter to exist; matter is necessary in order for time to exist; and time is necessary in order for space to exist. They are codependent.

So, if time is just some form of a dream, an illusion, as many philosophers have speculated, then so are space and matter. Yet from the standard or Copenhagen interpretation of quantum physics, we understand that matter cannot exist without an *observer* of matter.[1] Hence we are led to the notion that all four qualities of existence—space, time, matter, and mind—are codependent; they all arise simultaneously. In this chapter we will examine these ideas.

Observation requires time. In fact, observation plays a very special role in the nature of time. We could say that without observation or observers, time would cease to exist.

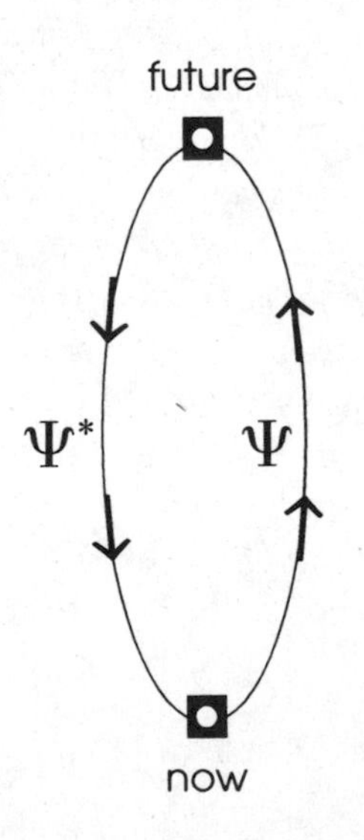

Figure 7.1. A Time Loop As Seen in Quantum Physics.

When quantum physicists determine the probability of an event, they calculate a number. This number arises from the multiplication of two mathematical functions called *quantum wave functions*—or, as I call them, *qwiffs*. Qwiffs are imagined to be real waves moving through space and time. However, they are not real waves; they are purely imaginal. They are not fields like magnetic fields or gravitational fields. They cannot be measured. They have neither mass nor energy. They exist in our minds and imaginations. That is, they do not exist as we observe real material things existing. Qwiffs' mathematical functions require us to use both real and imaginary numbers.[2] Usually these functions depend on time, that is, they specifically change as a function of a symbol, *t*, called the "time variable." We imagine the time variable to either increase or decrease in value and we determine how these functions change in either way. One of these functions is called ψ (Greek letter psi) and the other is called ψ* (psi-star): ψ represents a wave moving forward in time; ψ* represents a wave moving backward in time.[3]

We imagine qwiffs to exist everywhere in space and time. They take on specific roles between what are called *end points*, or *boundary points*, in space and time—the initial and final events of the history of an object. We think of these boundary points as real events and we picture the waves moving from one end point to the other in time with ψ going forward and ψ* going backward (see figure 7.1). The two qwiffs make a time loop. From the vantage point outside the loop a present moment connects in a single loop in time with either a future event or a past event.

The dynamic laws governing time loops bring a story into being. In other words, when a time loop is created, the world we commonly and uncommonly experience as "out there" arises both in our minds and in what we believe is objectively shared reality.

From time loops we can picture all of reality arising as the creation of a story. Once the end points are decided events, the story oozes out of the imaginal realm into the actuality of space and time. The story becomes a reality. A time loop emerges from the imaginal into the real as if it gushed out from each event. The story moves forward in time from the beginning event and backward in time from the final event. The story arises in between the two events.

These boundary events themselves are also not real until the story line emerges. One event does not *cause* the other and neither event *causes* the time loop. The end events themselves are just overlaps of possibilities in the imaginal realm. When they connect in the imaginal realm, and thus create a time loop, the whole story emerges at once as a created reality. The important point here is that the end points emerge into space and time because of our *desire* to have a story. You and I conceive of the beginning and ending events and then we measure them, see them, or sense them in any manner given to our senses and our abilities—our sensibilities. The story emerges at the instant we sense it, not before. We create both a past and a present or a future and a present for every story we believe to be true.

In the 1970s the popular idea "creating your own reality" arose as a New Age aphorism. Actually, the idea had its inception in quantum physics. Physicists, by this time, had realized that quantum physics *did not* predict actual events but *did* accurately predict probabilities for events.[4] But what determined the actual events? Whenever the predicted probability for an event, say event 1, was near unity (meaning it was highly likely to occur),[5] it tended to appear far more often than not when an experiment involving the possibility of event 1 was repeated many times. If event 1 and another event, say event 2, were predicted to occur with fifty percent probabilities, sure enough, the measured or observed results after many trials tended to be fifty percent 1 and fifty percent 2.[6]

The problem was, how should the predicted probability for events and the reality of those events be considered? Probabilities deal with the realm of the mind (the "in here"), and actualities deal with the realm of the senses (the "out there"). Since physics supposedly deals with "reality"—the reality supposedly "out there"—were these probabilities also "out there" in space and time? Or were they "in here" in our minds only? In other words, did the probabilities of quantum physics somehow seep from the imaginal realm onto real events? Did this field of mind stuff—possibilities—squeeze itself into space-time as if it were toothpaste from some invisible tube oozing out onto a toothbrush?[7]

As strange as this may seem, if this were the only problem with quantum physics, it would have been considered no more difficult than other physics theories as, for example, in statistical mechanics or in statistical thermodynamics where probabilities occur very naturally, with no oozing.[8] In these branches of science we never consider such questions. We know that events such as 1 and 2 occur because real objects *cause* them to occur and we are simply ignorant of these causes. The events in question occur, no doubt, "out there" as, for example, when you toss a coin and cover the coin with your hand before it is seen by anyone. The probabilities for event 1, heads, and event 2, tails, are certainly in your mind and in the minds of other potential ("in here") observers of the coin. Uncovering your hand will reveal one of those events, either 1 or 2. That uncovering certainly takes place "out there" in space and time.

But in quantum physics things don't work this way. It appears that events 1 and 2 are not always independent of our mindful influences on them.[9] For example, suppose we replace the coin by an electrically charged spinning particle such as an electron, and the coin-flipper by a magnetic field. Magnetic fields are imagined to have unseen lines in space called "field lines" that point from one pole of the magnet to the other. Magnetic fields flip charged spinning particles so that a spinning particle's spin axis points along the direction of the field lines or in the opposite direction, like a coin with heads or tails.

Here, event 1 becomes spin-parallel and event 2 becomes spin-anti-parallel—the only two possible observations that can be made of the electron's spin-axis direction with respect to the magnetic field's direction. We say the electron has *spin up* or *spin down.*

We may think this is not a problem. But quantum physics is a very strange theory. Suppose the magnetic field lines run from north to south and suppose we decide to do an experiment where we insert a sideways spin-east electron (a real electron with its spin axis pointing toward the east). Quantum physics says the *real* single spin-east electron will behave as if it were two *imaginary* electrons: a spin-north and a spin-south electron each in a parallel world. To illustrate, picture sending such a sideways (spin-east) electron into a north-south directed magnetic field apparatus. The device directs the electron's motion into an upper pathway if the electron is flipped into the spin-north state or a lower pathway if it is flipped spin-south. Then the electron is brought back along its original direction of motion so that it emerges from an exit hole moving in the same direction it had when it entered. As I said, the magnetic field device *should have* flipped the electron's spin one way or the other. Consequently, we *should* see the emerging electron with either spin-north or spin-south. But, as has been well proven in countless experiments, the electron never emerges from the device in a north or a south state but in the east state it originally had before it entered the device (see figure 7.2).

Now this is a mystery, if we think about it: since it seems that the electron should go along one pathway or the other, there should be no way for it to become spin-east again (except by a fluke). Again, this might take a little thinking. Look at figure 7.2 and consider the options. In the first option, the north-pointing

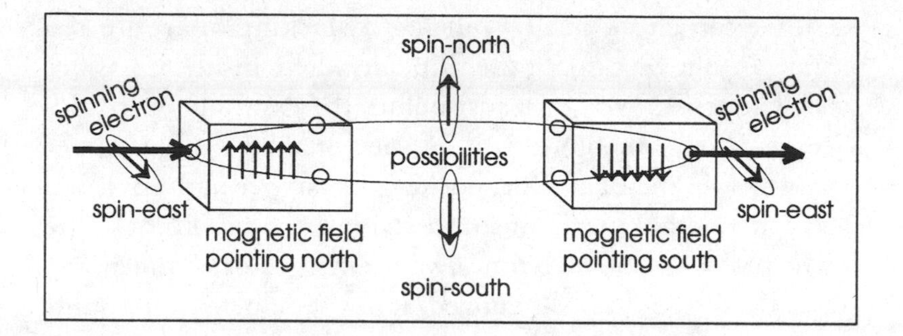

Figure 7.2. Parallel Realities Recombined. A spin-east electron in the Stern-Gerlach device (named after its inventors).

magnetic field should have flipped the spin from east to north so that the electron moves along the upper pathway. In the second option, the magnetic field should have flipped the spin from east to south so that the electron moves along the lower pathway. In either case, the magnetic field in the second box should merely confirm what the first magnetic field has done. It can't flip the spin into an east direction because the field points in the south direction. All it can do is confirm that the electron is spinning northward or southward. If the electron had spin-north it would continue to do so. If it had spin-south it would also continue to do so, as it finally exits the box.

So, we *should* expect to see the electron emerge with either a spin-north or a spin-south configuration. But it never does that. It always emerges with spin-east just as it had before it entered the two-box set-up. We can only conclude that somehow the electron cannot be represented by either pathway alone, but must be represented by both imagined pathways simultaneously. We imagine the electron to exist in two parallel worlds (spin-north and spin-south). These imaginal worlds overlap to make up this world and to give us the spin-east result we see. By the way, I do mean "imagined," for if we try to see what is going on, everything changes.

We can check this by putting a blocking filter into the device; in fact we can put the filter anywhere inside. Consider what happens if we put it just before the exit hole of the device, as shown in figure 7.3. If the filter blocks the upper pathway (see the black dot blocking the upper pathway), the electron emerges in a spin-south state. If the filter blocks the lower pathway, the electron always emerges in a spin-north state. But if the filter is removed, it always emerges in a spin-east state.

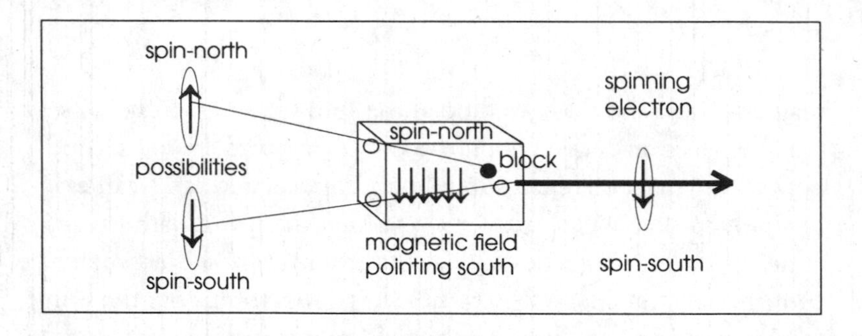

Figure 7.3. Parallel Realities Separated. A spin-north possibility is blocked in the Stern-Gerlach device at the last minute.

Which state is the electron really in just before it emerges from the exit hole? After all, hasn't the first magnetic field done its work and flipped the spin-east electron northward or southward? Suppose we put the block in one of the pathways (or not) just before the electron passes through the final exit hole. How can what we do here, after the electron has made its way through both boxes, bring about or change anything that was done to the electron earlier? Common sense regarding time says that nothing we do at the end of the experiment should affect the starting conditions of the experiment. But there's the surprise! What we do at the end does determine, once and for all, the complete history of the electron—including its beginning and its history. That history depends on us. We ultimately determine the north, south, or east spin direction it had after impinging on the first magnetic field by:

A. inserting the filter in the lower pathway at the last instance;

B. inserting the filter in the upper pathway at the last instance; or

C. not inserting the filter at the last instance.

Hence we create not only the reality of the electron's spin direction, but we also create its history. If we do A we know the electron emerges following the upper pathway. If we do B we know the electron emerges following the lower pathway. But, if we do C, then what? How do we think about the electron? Most physicists imagine the electron following both pathways simultaneously so that the images interfere with each other and thus reconstruct from their potential (imaginal) north and south states the observed east state. In any case, we create its reality, its historic inertia, the very definition of its own existence in space and time, and hence the slogan, "The observer creates the reality."

And so the story goes. We are seeing mind becoming matter.

Subjective Time, Observation, and Imagination

We have seen that a real history (the space-time trajectory) of an electron depends on both what we do at the start of the electron's "imagined" trajectory and what we do at the end of its course through a device. When and where do we become aware of such an imagined history; when does an imagined history become a real history? The facts seem to support this remarkable conclusion: Although a history depends on our observations of both the starting and finishing events, we remember the history as if we were aware of it while it was taking place.

In other words, we seem to "live" the history as it happens. We make it a "living" story. We live in a river of time in which the source of the river (our past) and its final destination ahead of us (our future) already exist. (I know this sounds crazy, but please bear with me a little while more as I take this one step farther.)

Notice I use the words *observed* and *imagined* here. The observation of an imagined history requires that both starting and ending events be specified. Leave one of them out and you do not specify a precise history; nor will you have a memory of it; nor will you experience it. Now, by "specify," I mean something rather specific: To specify a history, it must "enter consciousness," be "content for the mind," "lay down a track in memory," and so on. Since we all have subjective "in here" conscious experiences of the world we sense to be "out there," we all believe we know what these statements mean. But, it is difficult to find a testable mathematical or scientific model of conscious experience.[10]

Neurophysiological experimental evidence gathered by Benjamin Libet and his coworkers (see chapter 5) support the conclusion presented above. Nothing can enter consciousness until at least a pair of physical events occur and then a complete and logical causal history gets recorded *even if it never actually happened.* We can form memories of events, stories that make sense to us, so long as the imagined stories have acceptable end points. The story sets its own middle in place once the beginning and ending are set. Remember, a single event cannot be made conscious, it didn't really happen until another event confirms it. (And, there are several other interesting technical consequences that I don't have space or time to discuss here![11])

In Libet's experiments a person observed his or her own conscious responses "in here" to stimuli that were imposed from the outside, "out there." In some cases the "objective" measure of the timing of the stimulus events as observed by an "out there" observer (after the fact) were found to be time-reversed by the subject who was directly experiencing them "in here."

This reversal of "in here" subjective conscious experiences as compared to "out there" objective observations of those events eliciting conscious experiences may be what creates the "binding" of conscious experience as commonly sensed.[12] Subjective experience, because it requires two or more events, must necessarily become a story line, a history. As such, and only as such, will a history be recognized after the facts: first the end points and then the history. Indeed, the history emerges once the endpoints have been decided. But from the standpoint of beyond space and time—the imaginal realm of consciousness—we can just as well say, first the history and then the end points. Indeed, the endpoints emerge once the history has been decided. A complete "re-membering" of events takes place between the endpoints and is consistent with the end points.

Consider the concept of a personal history. Our memory trace or historical record, although it takes place in time, does not occur in space. Our minds, or consciousnesses, appear at the boundary between the real and the imaginal. The boundary between space and time arises here. Space is fundamentally and objectively "out there," while time is fundamentally and subjectively "in here." Yet they overlap. We imagine time as "out there" when we peer at a distant star and think about starlight taking many years to reach our eyes. We imagine space "in here" when we picture a scene of our past. Consequently, we become aware of a personal history before it ends and after it begins. But we report on that awareness, put it to memory, after the history is completed.[13]

Here we also have a new implication: In constructing a personal memory trace, we follow a logical course—one that ensures our species survival. In order to survive as a species we create consciousness as a collection of thoughts. These thoughts appear as pictures of causal chains of events. These events reflect our experiences or the stories we have learned from others. These experiences enter through our senses in physical processes. We usually don't remember bizarre things as really happening to us. We remember what we call a "classical" world, a rational world of cause and effect.

This is what consciousness does: It constructs a classical world[14] (apparently sensible, noncontradictory, and nonparadoxical) based on a quantum world (apparently nonsensible, contradictory, and paradoxical). Thus a memory trace appears as a fact-like classical story and not a quantum-like superposition of paradoxical and contradictory possibilities. We construct reality "out there" after the fact, with hindsight, but become aware of it as a story line, a history, even before it is completed.

In the above example of the spinning electron entering the unblocked box 2 (see figure 7.2), suppose we observe the electron at the end point of its trajectory and see that it is indeed spinning east. Remember the earlier question: What do we imagine about it in the middle, just after it emerges from the first box and before it enters the second box? Using a quantum physics picture, I drew it as two electrons, one spinning north and the other spinning south at the same time (as I have done in figure 7.2). Now, in an imaginal quantum world, this is not surprising. No problem, since in my imaginal world east isn't really east, it is north and south. One electron isn't really just one electron, it is two electrons. But in a real world, one that I observe, this cannot be. Only one electron shows up. Its spin must be either pointing south or north after emerging from box 1. In the imaginal world the electron spins north and south simultaneously. But in the real world this violates logical consistency. It cannot have these attributes simultaneously in any real world we live in. We will never see this happening. What's true for the real world isn't so for the imaginal world. In fact we must keep track of both spin-directions simultaneously to create the real and observed story.

Here a new idea emerges. A story is an imagined classical realization of a non-realizable—quantum-impossible—sequence of events. We create a world of events in our mind, giving to them just the attributes we require to complete the story, even though many of these attributes were never actually experienced. In other words, we fill in the blank spots (like filling in a vision where the blind spots in our eyes cannot see). These blank spots exist because of the uncertainty principle of quantum physics, which does not allow emergence of all of the possible details of the events. These various details cannot exist simultaneously in any single universe, but must exist as complementary possibilities in parallel worlds.

Go back to chapter 5 for a moment and look at figure 5.1. Remember this figure is an analogy of the quantum world. When we see the cube, we are faced with the cube's orientation relative to us as either looking up at its bottom or down on its top. These are the unique states of the cube as seen in a real world. But when we see the figure as black pieces against a white background, we are seeing a complementary vision. The complement also appears as a real world vision. But, we cannot see flat pieces and a three dimensional cube simultaneously. That picture exists only in the imaginal world. Seeing flat pieces means overlapping two possible "cube views."[15]

Thus the imaginal world is not just the visions of a disturbed mind or the wishful thinking of a hopeful person. We need it to make sense of the physical realm. Quantum possibilities transform into actualities: a world we all see and believe is "out there." Quantum physics teaches us that this happens because of a double flow within time; a source from the past sends out equally strong vibrating gushes (ψ) traveling forward in time, and a source from the future issues surging waves (ψ^*) traveling backward in time. The two streams intersect, and wherever they do, mind outside of space and time experiences what it's like to be inside of space and time.

Time loops like these double flows appear as stories. All stories must have beginnings and ends together with a connection between them. Stories arise as experience during the passage from beginning to end, even though the end has not occurred yet. Without the end already set in the future, the story does not arise. In this manner, the paradox of fate and free will is resolved. For along the journey, the mind must act continuously. Like a boatman steering a boat on a river, the boatman must direct his rudder, and he can do so in two complementary ways: going with the flow or set against it. Thus, our lives follow paths that we freely experience by using our abilities to choose from complementary possibilities (like the cube in figure 5.1). By altering our rudders, we set our minds hoping to make sense and find meaning for our actions in the simultaneously existing worlds of the "out there" and the "in here."

hhayt

ח

As we progress along our Hebrew line of letters, we see that each letter is flowing naturally out of the one preceding it. The eighth letter, hhayt, represents the fundamental synthesis of energy. With hhayt we have the dim beginning of physical existence. Although still a seed, hhayt represents the summing up or storage of the various possibilities represented by zayn.

Hhayt can be viewed as a pool of unstructured energy, not specifying any state of energy itself. But it is still unactualized, as such, because it is still in the seedling state.

Chapter 8

Meaning and Manifestation: The Great Gathering

There is neither spirit nor matter in the world; the "stuff of the universe" is spirit-matter. No other substance than this could produce the human molecule.

Pierre Teilhard de Chardin
Jesuit paleontologist

Einstein had a difficult time accepting the weirdness of the standard model of quantum physics. This model supposes that an observer has a strange effect on anything observed—namely the observer causes a single actuality to pop out from a cloud of possibilities. A pretty whimsical idea, difficult for physicists to accept as they like to explain things in a "rational" manner. So, to avoid this magical idea of a popping leap of mind into matter, one physicist invented the parallel universes—or many worlds—theory of quantum physics (as discussed in chapter 5). It says that the observer does not cause a popping action. Instead the observer merely interacts with the object being observed and if the object had several possible outcomes then the observer would also find him- or herself witnessing each possible outcome in a different parallel world.

Einstein once said to the inventor of this parallel universes theory, Princeton University physicist Hugh Everett,[1] that he couldn't imagine how a mouse could change the world simply by observing it. Everett replied that it wasn't the universe that was changed by the mouse, but the mouse that was changed by the universe.[2] Instead of pursuing a single course through history, the mouse follows a number of tracks in parallel universes—each track splitting off into a new comparable history as the mouse goes about sniffing his way, searching for the cheese. And, as goes the mouse, so go you and I.

Accordingly, each story line or history we follow *meaningfully* connects a single future event—an event that is destined to occur—and an actual past event. These events occur in each world and therefore exist in them all. If we think of the parallel worlds as sheets of paper, we can think of the endpoints as punched holes that go through every sheet. And our knowledge or present awareness depends on its relationship with both the one actual past event and the one destined future event that punches through all of the parallel realties in which we exist. Thus, what we experience now depends on both the past and the future in new and unexpected ways. So although it may seem to be so, a single, supposedly already established, past event, or even a sequence of past events, does not determine the present. The *future* also plays a role. This dependency of our present awareness on a single future event, a single past event, and the presence of other parallel story lines, not only establishes the story line we follow, it gives us the feeling of free will—the sense that we could have taken another path through life. In a manner of speaking, we did.

Yet life seems to go by so ordinarily. Why can't we become aware of these other histories? Why don't we see into the future? What makes us singly aware of this story line and where are those other *us*es anyway?

As it turns out, the story line with the greatest probability becomes what we are aware of. It becomes our remembered history. This history, in turn, becomes significantly meaningful. And meaning itself has a quantum physical basis: Significant meaning arises as the most probable story between an event in the future and an event in the past. Hence, that which is meaningful turns out to be that which is most probable, the most plausible, and that depends on a lot of things including what we believe and what culture we accept as real.

Parallel universes—other worlds inexorably linked to our own—play a critical role in the functioning of these story lines. When we see things anew, we enter a parallel universe where we have the opportunity to reinvent ourselves. In so doing, we enter a new parallel world. We already exist in these parallel worlds; only our viewpoint has changed. The search for meaning in life is actually a real search—an odyssey or exploration—into multiple parallel worlds of possibilities that appear to change with each and every action we take. Our observations really don't change us or the many worlds around us, but they do provide new meanings in our lives when we shift our viewpoint between worlds. This scheme works because extending around each and every thing is an alchemical aura in space and time. This field is quite physical. It appears as light and this light is not captured or confined to any single world or event, but corresponds to transitions *between* worlds or events.

Joe Sixpack, A Noh Drama, and Any Dictator You Care to Remember

Picture yourself taking a stroll in your neighborhood. As you meander down the street on a warm summer evening, you pass a bar where patrons are spilling out onto the sidewalk. You decide to pay attention to one of the patrons, a character we shall call Joe Sixpack.

Joe is drinking from a bottle of beer and talking loudly. He takes note of you as you pass by. Perhaps you are an attractive woman, and it is not a surprise that Joe would notice you.

Perhaps you are a middle-aged man dressed in jeans and a sweatshirt, and as such you don't command any attention from Mr. Sixpack. Or perhaps you are wearing a suit and tie, or a tuxedo, or a wedding dress. Imagine the look on Joe's face as you walk by in each of these scenarios.

Now I want you to imagine that surrounding you is a giant bubble. You are in the bubble's exact center, and thus it extends in all directions from your body, even into the ground you walk on and into the air above your head. You can call this bubble your *aura field*, if you like. The bubble forms your field of alchemical possibility. It represents how far from your body your presence is detectable, both in space and—as your image exists and persists in people's memories—in time. The bubble also extends into parallel worlds. You may be detected, therefore, in many different ways. People can see you from a distance. They can hear you coming. Perhaps they can even smell you coming—you have "worked out" recently and haven't bathed, or you are wearing perfume. Or, perhaps someone is thinking about you, remembering you as a child; your field extends far back in time to when that person encountered your younger self.

Now look at Joe. As you approached the bar, you saw him from a distance, and since he was drinking and loud, you may have heard him well before you saw him. Joe's bubble extended out from his body as far away as the distance you were when you first sensed him. You shouldn't be surprised, then, that he and others in front of the bar also *sensed* your coming. It's not important to spell out which of the six senses were used to detect your presence.[3] It *is* important that you were sensed well before you actually walked in front of the bar. Your field goes with you wherever you go. And it changes. Sometimes it extends quite far, a hundred meters or so, particularly on days when you wish to "announce" yourself to the world. At other times it may extend only a few inches around your body. On these occasions perhaps you are dressed down, feeling tired, or just plain wish not to be noticed.

Japanese Noh dramas provide an interesting example of aura fields in action. Noh players wear various costumes as they appear on stage. Those commanding the most attention are the most colorfully dressed—sometimes having great manes of orange hair, suggesting that they are lions, and wearing elaborate, expressionless white masks. As these characters move around the stage, they often dance or leap.

As the drama proceeds, music is played by "invisible" musicians—they can be seen on the stage only if we look for them. Sometimes the music is sad and causes us to "see" sadness in the great white face of the orange-maned actor. At other times, the invisible musicians play joyful music as the actor leaps, and now we "see" happiness. The expressionless white face suddenly seems to be smiling. We might think that the white facemask changed when we weren't looking. Perhaps. But more likely it didn't. We were simply experiencing the aural field of the actor. It extends quite far into the space and time of that theatrical experience. It includes our feelings as we respond to the actor. For this actor is to be noticed; he is the show. But without us no show exists. Therefore, we can change the show.

For example, from time to time our attention will shift and suddenly we see other people on stage, often smaller in stature, dressed completely in black with their faces covered. These people move stage props and change the backdrop scene. When I first saw a Noh drama, I was instructed to ignore these people in black. Of course as soon as I was told this, my attention went directly to them. In fact, it took me a while to begin to "forget" them, a task that became easier as the drama developed and I became highly interested in the orange-maned actor.

It is easy to consider actors as having large aura fields. Screen actors and actresses have huge auras—consider the late Audrey Hepburn, or Sean Connery, or Madonna. And not all aura fields are amusing or pleasant. Notorious people have large auras as well—consider Saddam Hussein, for instance.

The dictator's field is all over Iraq, in every home, in every farm, in every factory. His image, which is constantly changing through the use of different photographs of his face and body, is never allowed to vanish from the minds and memories of the Iraqi people. In fact, all dictators use this technique. Their fields extend over wide ranges of space-time—particularly time, as their images persist in our memories.

We may ask ourselves where this aural field exists. In the people themselves? In our minds as we observe them?

Well, continue on your journey through your neighborhood thinking about Joe Sixpack, the Japanese Noh players, and Saddam. As you do so you may get the idea that it takes two to make an aura field: the source of the field and the observer, or detector, of the field. Just the source by itself, although it may be emitting this field, broadcasting it in space and time, does not make a field. If there is no response from the possible observers, or the audience, the emitter has no way of knowing that his or her field is even present. When actors get no feedback from their audience, they may alter their performance to "adjust" the strength of the field. Dictators may enforce more or less terror to strengthen their field. It all depends on the perceiver of the field.

Now, I am going to ask you to do something that will seem even stranger than the steps preceding this. Imagine that you are able to stop time. Do you remember playing with a flipbook when you were a kid? Flipbooks are small books containing a number of pictures on a single side of each page, one following the other. The pictures are all slightly different. You grasp the book's spine in your hand and then flip the pictures. What you see is the equivalent of a motion picture—single frames giving the illusion of motion.

Now, imagine the following in your time-frozen state. The universe consists of gigantic flipbooks—an infinite number of books and an infinite number of movie frames in each book. Think of this as a virtual library of single frames that exist in two dimensions, like pages of picture books. The key is that all of this is just data, bits or organized patterns, but when grouped and arranged—flipped, so to speak—they turn into the moving experiences of our daily lives.

The Timeless as the Seat of Time

Imagine a magician descending a staircase. With each step the magician descends, he releases potential energy in the same way a ball does when it falls down steps, bouncing its way to the bottom. Imagine that one unit of energy is released with each downward step. Now, imagine that the descending steps are numbered 1, 2, 3, and so on, with the number 1 corresponding to the top step, and the magician is standing on step 1. At some unpredictable time the magician will make a quantum leap—a transition to a lower step. He could leap down to the second, the third, the fourth . . . or the ground step. The actual number of leaps depends on how many steps there are to the basement.

Let's suppose the stairwell has six steps. No matter how many leaps he takes, as long as the magician reaches the bottom step, he will release the same amount of potential energy. In fact, he could even descend a few steps, then climb back up a step or two, and descend again. He will always release the same amount of energy so long as he begins at the top step and finishes at the bottom. Thus, there are many ways he could make his way to the last step.

In table 1, I have listed some typical downward paths wherein the magician never attempts to reclimb any step. The table shows the different number of ways a descent could take place depending on the number of stairs and the number of leaps taken. For example, descending six stairs the magician could either leap directly to the bottom or he could make his way down one step at a time. The most likely way, however, has him descend by taking three leaps. He could do this in six different ways. Thus if we don't consider *how* he does it, but only count the number of leaps he takes, it becomes most probable that he will descend in three leaps.

As goes the magician, so goes the whole universe. For it appears that the universe is also following a downward path of a certain kind. It appears to be descending, only this time the descent is in time and the downward spiral appears to be a gain of information and a loss of order—the universe appears to be gradually getting smarter as it also appears to be losing order and becoming more chaotic. The question is, how? The answer is, again, in leaps. And it appears to be following all possible paths downward. The most likely sequence appears to be the only sequence followed; but in reality all of them are taking place simultaneously in alternate universe-possibilities.

Yet there is something constant in all of these universes: the total energy released. Given that the beginning step and the final step are determined ahead of time (they may be the moment of the big bang and some appropriate ending moment of time, possibly the big crunch); given that the energy released along each step is the same; then the final tally is always the same regardless of the story and regardless of the likelihood of the story being true. The universe appears to conserve energy as it becomes more chaotic and more information-rich.

Picture these alternate stories as pages of the flipbook.

Imagine being in a typical three-leap story in a six-stair descent. For example, you leap from steps 1 to 3 to 4 to 6. In doing this you give up two, then one, then two energy units, five in all, as you make your way down. All of the other story possibilities are also present and the total energy released is the same in each story. But the energy released in those other stories will have a different distribution. For example, in a two-leap story, say 1 to 2 to 6, the energy will be one unit then four units of energy, again five in all. Although you are not aware of the other story possibilities, nevertheless their energy releases will be sensed in your

Table 8.1. Number of Different Possibilities.

Number of Steps in a Story	Sequences of Steps	Number of Leaps	Energy	Number of Stories
2	1, 2	1	1	1
3	1, 2, 3	2	2	1
	1, 3	1	2	1
4	1, 2, 3, 4	3	3	1
	1, 2, 4	2	3	2
	1, 3, 4	2	3	
	1, 4	1	3	1
5	1, 2, 3, 4, 5	4	4	1
	1, 2, 3, 5	3	4	3
	1, 2, 4, 5	3	4	
	1, 3, 4, 5	3	4	
	1, 2, 5	2	4	3
	1, 3, 5	2	4	
	1, 4, 5	2	4	
	1, 5	1	4	1
6	1, 2, 3, 4, 5, 6	5	5	1
	1, 2, 3, 4, 6	4	5	4
	1, 2, 3, 5, 6	4	5	
	1, 2, 4, 5, 6	4	5	
	1, 3, 4, 5, 6	4	5	
	1, 2, 3, 6	3	5	6
	1, 2, 4, 6	3	5	
	1, 2, 5, 6	3	5	
	1, 3, 4, 6	3	5	
	1, 3, 5, 6	3	5	
	1, 4, 5, 6	3	5	
	1, 2, 6	2	5	4
	1, 3, 6	2	5	
	1, 4, 6	2	5	
	1, 5, 6	2	5	
	1, 6	1	5	1

The last column tells us how likely it is that the particular scenario can take place. For example, in the six-step story a three-step leap is six times as likely to occur as a single-step leap or a five-step leap.

universe as an aura field of possibilities—a light-glow or sense of light presence. You won't see this energy or feel it consciously because its total always amounts to the same total you are experiencing. You *are* aware of the usual or normal total of light energy; you are *not* aware of the unusual or abnormal total of light energy.[4] The key here is that this abnormal energy will nevertheless exist as light, a glow, an aura in your mind, but will not be contained in the flipbook of the material universe. The aura has physical consequences, yet it is not physical.

Sense this light in your mind and you are no longer stuck in this universe. You transcend your universe and sense the way of God's working. Science can help us understand this, but it is a concept not fully comprehensible through the intellect alone. It must be recognized through experience. And this is not easy. It is best done in silence and humility. There is no way to do this on one's own. One must surrender to a timeless sense of higher purpose, to a sense that there is a consciousness beyond one's own grasp of this consciousness.

Can such recognition occur? If it could, you would be beyond time and space. And without time, everything would be present in an eternal *now*. Light moves in eternal now.[5] Thus, if you were to see yourself as light, time would cease to exist, and you would experience space vanishing. This experience is true for all light radiation, even the light within our own bodies traveling from one place to another as it mediates the many processes taking place inside of us.

Mind also contains a sacred light. It, too, moves without experiencing space or time. We can perceive this light within our own mind. It exists in the imaginal realm of our essential and subjective being. We can see it only by looking within. We all know secretly that it is real. We should, for we *are* that light.

tayt

ט

With the letter tayt we come closer to existence than with any other letter so far, for it is the result of hhayt—a direct formation of the concept of "come together." Tayt represents the cell—any focus or center or concentration of energy—that becomes female.

In a sense, tayt is similar to bayt and hhayt in its action, but it is more defined than they. It is any focus, center, or concentration of energy, female in its action. It is a further summing up of the wave function represented by hhayt, in a special way giving rise to a primary structure of a cell or a womb, a place for birth to begin.

CHAPTER 9

Structure and Beauty: Spirit and Soul

The dual substance of Christ, the yearning so human, so superhuman, of man to attain God . . . has always been a deep inscrutable mystery to me. My principle anguish and source of all my joys and sorrows from my youth onward has been the incessant, merciless battle between the spirit and the flesh . . . and my soul is the arena where these two armies have clashed and met.

Nikos Kazantzakis
The Last Temptation of Christ

In this final chapter we will look at the notion of spirit and soul from the perspective of new alchemy. A key insight comes from the idea that matter and spirit are in conflict with each other. Hence, to work through this conflict we need to look to the ultimate transformation: the evolution of the human being from the Adam Kadmon, its original source.

A lifetime is a strange journey. It is a round trip. We end up where we began. Remember that Adam Kadmon in Hebrew, אדם קדמון—read from right to left, aleph-dallet-mem qof-dallet-mem-vav-nun—has a symbolic meaning. Aleph, the primordial spirit, acting through the idea of resistance, transforms this resistance of consciousness into cosmic possibilities. As it turns out, dallet-mem, דם, can also be written, דמ. Notice, dallet doesn't change, but mem does. It can have two forms: a final form, ם, and a regular form, מ. It is the same letter, but its significance changes. Mem-final represents a cosmic or final ending (an outcome of great possibility), while mem-regular represents an evolved resistance—referring literally to the water that makes up eighty-five percent of the brain; this could mean that without water our brains could not function as memory devices and that without the resistance that water offers to the brain's electrical activity, consciousness would not occur.

One more insight into the meaning of Adam: dallet-mem means *blood* in Hebrew. Adam, therefore, means *spirit in blood*.

So, in relation to our round-trip concept, Kadmon suggests something very interesting. Qof-dallet-mem, the *kad-mem* of Kadmon, is just like aleph-dallet-mem when we consider that "qof" is the exalted form of "aleph." Thus we see that when aleph and yod, spirit and existence, reconcile (qof) in the blood (dallet-mem), endless creation and fertilization (vav) of the fully cosmic and evolved human (nun-final) occurs. In brief, we return to God; we experience the full realization and recognition that we *are* Adam Kadmon. We have come full circle; we have made a round trip.

Without this recognition of one's spiritual nature, one's soul, one's Adam Kadmon potential, the alchemical journey is not only a frightening one to take, it becomes heartless and strangely meaningless. With the recognition of one's spiritual nature, however, the journey becomes heartfelt and beautiful to behold. This beholding not only occurs "out there," using the light we can see with our eyes, but "in here," using the light we can see with our mind's eyes—our imaginal realm searchlights. I would call the ability to behold this splendor the complete evolution of the soul and the revelation of the meaning of each and everyone's existence in matter. This was the spiritual secret of ancient alchemy and is the basis of the new alchemy presented in this book. Thus if we remain stuck in the paradigm of Darwinian evolution and natural selection, we are destined to believe that evolution is purely accidental and that beauty and purpose are somehow evolving out of a natural selection from a random mélange of carbon, hydrogen, goo, and chaos. A pretty dismal picture. Personally, I can't fathom what life would be like without hope, beauty, and purpose. I can't see much purpose in survival for its own sake.

So, in the view presented in this book, I have attempted to show that the very capacity to tune to a finer vibration of the soul and to arise to a cosmic understanding of one's self defines something essential about us. To begin with, as the ancient alchemists did, I use, as models, metaphors based on my understanding of how the physical world works. Using metaphors allows me to explain things that are unfamiliar in terms that are familiar. For example, the concepts of soul, matter, spirit, self, and consciousness can be defined by imagining two basic physical objects: one is a vibrating string as you might see on a violin; the other is a mirror which reflects back images of the real world. In my model, both are placed in the context of quantum theories. Spirit would be akin to the vibrations of the string. We can imagine that the string is infinitely long and, due to random inputs of heat, air, or just the fluctuating vacuum of space, it vibrates. Its vibration represents the movement of the spirit.

This constant movement of energy—or life—provides the *modus operandi* of the string or spirit. In modern science, physicists understand that we can model the vacuum as if it were filled with an infinite number of vibrating strings, and thus the vacuum itself becomes vibratory and a natural place to look for a metaphor for spirit. The soul, then, appears as the reflected vibrations of the vacuum within the domain of time. The soul (and time) extends from the beginning and ending points of time, known respectively as the big bang and the big crunch. The vibration reflects from these end points, just as an image reflects from a mirror. The reflection from the ends of time gives the soul consciousness in the same way that we become self-conscious when we gaze at our own mirror reflection. Our mirror-image consciousnesses arise in space. The soul consciousness arises in time. The soul embodies in matter as the "self," or the "self-process." The soul relates to itself continually in the body and therefore it involves itself with the survival of the body. The soul isn't necessarily embodied to begin with, but the self must be.

Consciousness or mind in general arises whenever a reflection in time occurs. That means something reflects from points in the present or the future and the past, or even both past and present, or future and present. What reflects depends on the form of the consciousness. If we're talking about primal reflections from the beginning and ending of time, then the reflection produces a conscious and cosmic soul. I would call this the one Soul that inhabits each and every being.

When reflections are from points in space, then those become essentially unconscious chunks of matter.[1] Thus time forms the arena of consciousness, and space forms the arena of unconsciousness. Of course the distinction between consciousness and unconsciousness is more complex than this, since space and time fuse together according to relativity theory and the behavior of light.

The self, because it reflects a consciousness (the soul) in unconscious matter, contains both elements. So, the self behaves both consciously and unconsciously. Being alive and continually processing information makes the descriptions of soul and self difficult. So, the challenge we face is to define the *processes* rather than the entities themselves.

The alive and vibrant soul subjectively experiences life through our bodies. The world that we see with our everyday eyes—through the filter of our senses—derives from a more "objective" world. That "out there" objective world and the subjectively experienced "soul world" conflict with each other. Spiritual teachers have taught us that when living spirit descends into objective matter a fight ensues. So if we become too involved with the objective, external processes of life, we tend to lose touch with perception from the soul level. When we go within to an internal quietness, as in meditation, we begin to perceive something deeper and more meaningful than just the objective "out there-ness." So, if we have lost touch with our souls, we need to spend some quiet time—not in thinking, not in going over the day's list of everything that has to be done, but in being with ourselves in ways that allow a deeper inner reality to bubble up from within our consciousnesses.

Now on a human level, some people view their lives and think: "Oh God, what did I live for? Isn't it terrible that I'm going to die? Life was black when it started; life was bleak when I was here; and it's going to be black again when life ends! Black to bleak to black again. Oh, God, what's it all about?" In my view, this blackness and despair has been designed into God's system. We may not completely believe or even remember this design in this moment, but we have actually created all of it. The "me" that created it is not the person, the personality that identifies itself as Fred, or Martha, or Sam—that's not that person I am speaking to or about. It's the greater essence of *I*, this deeper presence, the Adam Kadmon within, the working of consciousness itself that is in me, in you, in everyone. That *I*, working through this body, is the same *I* that is reflected in the archetypal images of Jesus, Moses, Mohammed, Krishna, all of whom reminded us, and continue to remind us, of our true essence. These beings are reflections or representations of our own identification with our greater, deeper *I* self. They are the unbroken containers of the original light, which shone from the eyes, ears, nose, and mouth of Adam Kadmon.

The Light Which Cannot Be Contained

Ordinary matter possesses inertia—it resists our every push or pull; and when we do push or pull it, we feel a push or pull upon ourselves. Newton called this *action and reaction.* Light too, can push or pull on matter, but light particles have no inertial property. We can see light push or pull on objects (as in the case of a laser cutting a piece of steel or mending a detached retina), but we can't see any objects push or pull on light. Physicists call light's inability to be pushed or pulled its *absence of any rest mass*—rest mass is really just another way of saying inertial mass, the mass an object has even if it just sits there and minds its own business. Light will have nothing to do with being at rest. It's always on the move.

Matter, the kind that resists our pushing and pulling, likes to be at rest or at least not changing in its direction or speed. It maintains the status quo as well as anything else can. Matter is represented by quantum waves that express where and when you are likely to find it. These waves are invisible and can perhaps be thought of as undulating guiding patterns. These patterns, however, do not really exist in space and time; they are purely imaginal forms, although we commonly tend to think of them as if they were actually out there in space and rippling through time.

I like to think of these waves as purely imaginal ordering rules that govern the multiple physical, material, and inertial realities we call parallel universes. These rules manifest as waves of imagination in any one universe, but being rules they aren't really waves at all. When we relate what we imagine to be occurring in one universe to what we imagine to be occurring in other universes, the pattern in the pictures we bring to mind looks like waves. Think of the universes as if they were cards in a deck. Imagine spreading the deck out on a table in a pattern (like a typical Tarot pattern or a game of solitaire). We can think of matter as imagined dots printed on each card-universe and the waves as the spread patterns of these card-universes.

When it comes to light, the wave patterns governing light waves and the light waves themselves are exactly the same things. When we see light, we really don't see light at all: we see an effect appearing as a result of light pushing and pulling on the matter making up our sensory bodies. We see matter moving. Light itself is really out of this world and, as far as I can tell, out of any parallel world we wish to think about.

In the previous chapter we saw how a simple act of leaping down a flight of stairs can occur in a number of different ways. We saw in the example of the six-step flight that the most probable leap the magician would take was three steps. This was the most likely journey in that there were six ways to skip down those stairs in three leaps, more than in any other way. Hence if we were to actually leap down the flight of stairs without taking note of how we did it or how many steps we took, we most likely would do it in three leaps. From a parallel worlds way of thinking, the six different three-leap scenarios overlap and constitute a memory in which the differences in the leaps, the details of which steps were actually taken, are fuzzy or simply not recorded in memory.

Most memories are like that. We remember something, but quite often details are unclear. John said this and I did that; or, no, John did that and I said this. Reality is a construction of what we remember. We might say that the world consists of happenings, fuzzy events, or we might call them partially-real events. When we attempt to determine just exactly what those events were, in effect we create them as memories. We create a past and at the same time, depending on the results of what we remember, alter ourselves by redefining our expectations of the future.

Let's go back to our leaping-magician example. When these leaps take place, light energy is expressed. But since this energy is not inertial—matter can't push or pull on it—it really doesn't attach itself to any parallel universe. In the flipped-pages analogy, light would be our fingers flipping the pages while the matter being flipped would be the pages themselves.

Given that the same total light energy is expressed in every possible world, the only difference in each world would be in how that energy is distributed. The most likely distribution would consist of three photons—quanta—"particles" of light released in the three leaps.

Now picture yourself remembering a typical three-leap sequence, an average occurrence. Deep within your mind lie the other sequences, the ones that weren't as likely. For example, the sequence where the magician descends the stairs in one leap is in memory, but its memory is overwhelmed by the six overlapping three-leap memories. Exercising your ability to go into silence through meditation, to quiet your mind, opens up the possibility of recalling these other less probable memories.

If you do this while you are awake and actively seeing the ordinary three-leap world, you will have an extraordinary experience of seeing auras. These auras constitute a field within the mind consisting of the nonordinary sequences of possibilities and they are not seen by most people simply because they are not looking for them. Auras form the afterglow of alternate realities—an attempt of ordinary mind to see, as God sees, that everything glows. This glow or aura proves the reality of God because we, for a brief moment, see from a point of view that is not our own.

No one really knows how inertia arises. I'll suggest that the quantum physical formula for existence and the reason for structure and beauty arise from parallel universes—worlds not ordinarily experienced. Structure and material inertia arise in the overlapping of the most likely sequences of events. Inertial matter comes from the overlap of many nearly alike parallel worlds. The more alike the world sequences of an action appear, the greater that sequence resists change. Thus inertia arises as a consequence of similarity. Beauty arises as the difference from ordinariness. Experienced in mind as a glow or aura from new sequences—new possibilities—a new distinction separates from the ordinary or most probable distinction or perception. Hence beauty and resistance, the kind spoken about by Kazantzakis in the opening quote, naturally go together. Like all duality, one requires the other.

Awakening the New Vision

New alchemy takes place at the frontier of the imaginal/real realm. Our personal experiments in new alchemy produce new information and new transformational possibilities entering our dreams and, possibly more importantly, our waking thoughts. We awaken to a new vision of life and time. New meaning alters and changes what we believe and hence what we experience physically in the world, both personally and globally.

The new alchemy gives us a new and ancient vision of mind, body, spirit, and soul, and a new understanding of how the forces of purpose, creation, and transformation within each of us, when used consciously, can enhance the meaningfulness of everyday life. The possibilities are unlimited.

As I ponder my own existence, I see that as a species we often resist change and certainly transformational possibilities presented to us. Perhaps we seek to simply exist on automatic pilot, not having to deal with anything new. Passing through my sixties, I see that I have often resisted transformational possibilities when they have presented themselves to me. I can't tell you that every opportunity passed over necessarily led to some personal disaster in my life, but I can tell you that when I took advantage of a possibility and turned it into a reality, it always opened a new vision of myself, my relationship with the world, and the people closest to me.

Becoming something, turning a vision into an actuality, is not always easy. I hope that you have found in the pages of this book a little guidance in turning your own dreams into realities.

yod

י

Yod represents the world of structures and is the opposite of aleph. Thus where aleph was potential or able-to-be, yod symbolizes being. Where aleph was timeless and beyond any limit, yod is in time and limited.

In a sense yod, being aleph in existence, exists in a state of perpetual contradiction, for paradoxical aleph appears timeless and not existing and yet not non-existent, while yod is time-like and in existence. For as we become aware of existence we change it.

APPENDIX

Existence: Mind in a Box

I want to know God's thoughts, the rest are details.

Albert Einstein

In contrast to the Newtonian or classical laws of motion, quantum physical laws do not allow us to determine actualities. Instead, we face possibilities.[1] For example, using classical physics, when you flip a coin and allow it to land, you can predict with certainty which side will land face-up. To actually do this, however, you would find using classical physics difficult because of complications such as air currents, weight imbalances in the coin, and so on. Instead, you assign a probability of fifty percent that it will land heads, and fifty percent that it will land tails. You give up any hope of control. But within quantum physics we find a new and apparently paradoxical concept: separate possibilities can combine or interfere with each other, which leads to new possibilities. For example, the two possibilities of observing a flipped coin's face could combine and result in the coin standing on its edge. Usually we don't see this, of course, because the coin interacts with many things before it lands. But nevertheless, the separate possibilities of heads and tails, even though they do not manifest simultaneously, can actually interfere with each other and produce new possibilities.

Also, according to standard reasoning in quantum physics, a possibility is only realized when it is observed. A quantum object capable of existing in one of two possible states "suddenly" leaps into one of those states at the instant of observation. This is called the *observer effect.*

Consider again the example shown earlier in chapter 5 and repeated here. Perhaps this example is just a metaphor; or perhaps it is an actual quantum phenomenon involving quantum processes in your retina, nervous system, and brain. To participate with the quantum process, first observe the white box in figure A.1. Note the two distinctly front and rear square faces of the box. Now look again. Which of these square faces is in front? Is the lower-square face in front? Or the upper-square face? Or perhaps the box is leaping from the former to the latter? Or perhaps it is not seen as a box at all. Note the gray geometrical pieces pasted on the white background. Is it still a box? Why or why not?

Figure A.1. A Cube or Gray Pieces?

According to what has been called the Bohr or Copenhagen interpretation (named after Niels Bohr, a major contributor to the discovery of quantum physics) the box only becomes a real box, with one side apparently forward, when you observe it. In the Bohr interpretation of quantum physics, you cause the event (such as seeing a specific view of the box) to occur. Bohr said that an act of observation involved a merger of two ways of seeing: the classical world of perception and the atomic world of quantum events. By attributing the effect of an observer to an intrinsic eventual merger of the atomic world with the classical world, the question of the nature of the action of observation is left open. The Bohr interpretation says that when an observation takes place the property of the object under scrutiny mysteriously takes on a value. That is to say, we really don't know how an act of observation takes place.

Another interpretation, however, says that the observation of a physical attribute is nothing special and, in fact, when you see the box with a particular face in front, you also see the box with the other face in front at the same time—except that you as a second observer do not exist in this world but in a parallel world, or parallel universe! This has come to be known as the many-worlds interpretation of quantum physics.

How do you prove that the other world exists? According to the many-worlds interpretation, looking at the box once again, you see the superposition of the box simultaneously in its two position-states when you see the picture as gray pieces. Both position-states cannot exist in a single world. But the overlap can. The two states interfere with each other, producing in your mind a picture of gray pieces instead of a box with either side in front. In other words, you see both worlds at the same time, provided you do not see a box. If you choose to see a box, you do not see the gray pieces.

The many-worlds interpretation example given above visually illustrates this interpretation. Hugh Everett III, a graduate student at Princeton University studying under the highly regarded physicist John Archibald Wheeler, came up with the rather strange notion that even though quantum physics posits a view of the world that contradicts common sense, we should take it seriously.[2] If it says that two alternative possibilities can interfere with each other, overlapping together in one world, then somehow those alternatives must both exist simultaneously in other worlds. If possibilities affect each other by overlapping each other; if two or more possibilities somehow "add up" to produce a single-world (a consistent picture of reality); then somehow these possibilities must each manifest somewhere else. They must both be real.

Accordingly the two possibilities—although describing separate "you"s, each with a single perspective of the box, and separate boxes—really exist somehow in separate universes. Just as an "interference" pattern can be seen on a bright day when looking through overlapping layers of a fine mesh window curtain, the interference of both universes can be seen as the gray-pieces view. Only when a perspective view of the box was taken would each of the separate and parallel universes be observed.

In the many-worlds interpretation, when an observer observes the box, he or she, in effect, interacts with the box in a completely quantum physically predictable way and is thereby changed by the observation as well. If the box, after the interaction, exists in parallel universes, then so does the observer. Thus the many-worlds interpretation also explains the observer effect—the effect that an observer has upon a physical system simply because he or she observes it. Nothing magical happens when an observation occurs. The observer simply becomes part of the universe or universes in which the observation takes place.

As we have seen, paradoxical boxes possess an attribute called *perspective.* With respect to any given spatial direction, the perspective of a box can be seen as either looking

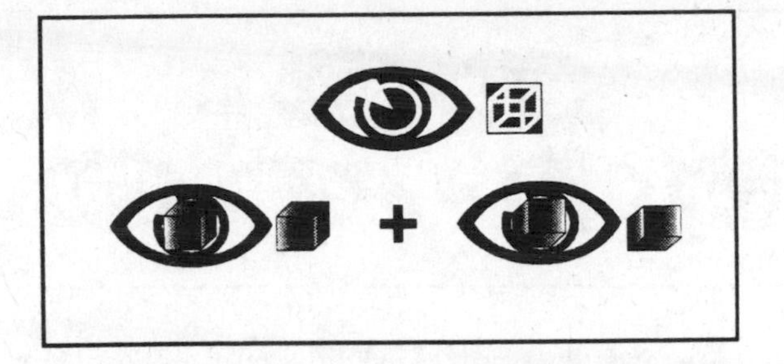

Figure A.2. The Cube and the Eye of the Beholder. The black-eyed observer at the top of the picture comes into contact with the mysterious paradoxical gray pieces/cube. By observing the figure as a cube he or she enters two parallel worlds.

up or looking down upon the box. Thus, the box can have perspective-up or perspective-down position. According to the Bohr interpretation, when you observe the direction of the box's perspective, it will instantly quantum leap into a state of perspective-up or perspective-down. The other perspective view vanishes in some mysterious way. But this doesn't happen in the many-worlds interpretation.

In the many-worlds interpretation, no leap occurs. The box splits into two parallel boxes (figure A.2) and the observer of that box (symbolized by the large black eye), or at least the mind of the observer that is able to make the distinction, also splits into two parallel minds. An observer of the box's perspective direction, and a box with its perspective appropriately matching the observer's observation, exists in each universe.

Look at figure A.3. You see what happens when another observer (symbolized by the gray eye) enters the scene.[3] If this observer also looks at the box, her mind will split. However, her mind will join with the mind of the first observer (the black-eyed observer) and they both will agree on what they see.

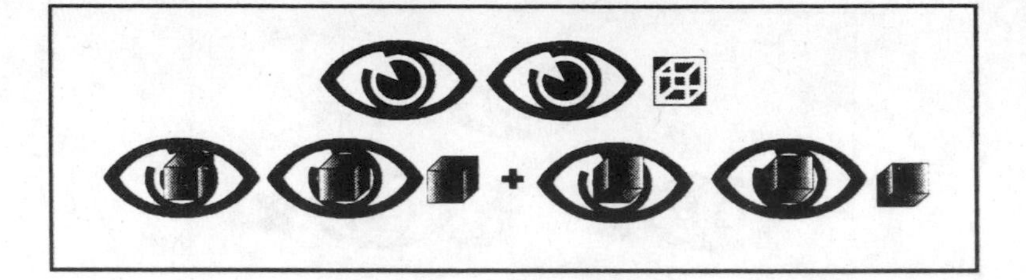

Figure A.3. A Second Observer Is Caught in Parallel Worlds. The gray-eyed observer and the black-eyed observer, at the top of the picture, come in contact with the mysterious paradoxical gray pieces/cube. By observing the figure as a cube, they both enter two parallel worlds.

If a third observer were to come along, the same thing would happen to him. In this way, what we call *consensus reality* is constructed and the interference between the possible parallel worlds diminishes.

After *N* observations by *N* observers (where *N* is an arbitrarily large number; see figure A.4, where *N=5* for simplicity), there will only be two well-separated universes. In each universe there will be *N* observers all agreeing that they saw the same thing (perspective-up in universe 1 and perspective-down in universe 2) and normally the two universes should no longer interfere with each other because of the correlation arising between all of the observers. In order to create interference between these two universes, each of the observers as well as the box would need to be overlapped so that each observer sees gray pieces.

Observers, however, are so complex that it becomes very difficult to create an overlap of them all. Consider for a moment that you contain all *N* observers in your head. Notice that you are now seeing both sets of your selves in parallel worlds. Now you can see why each observer only sees one perspective at a time. If you were in just one of these two worlds, the other perspective—the other "you" consisting of *N* observers all existing in a parallel world—would be unknown to you.

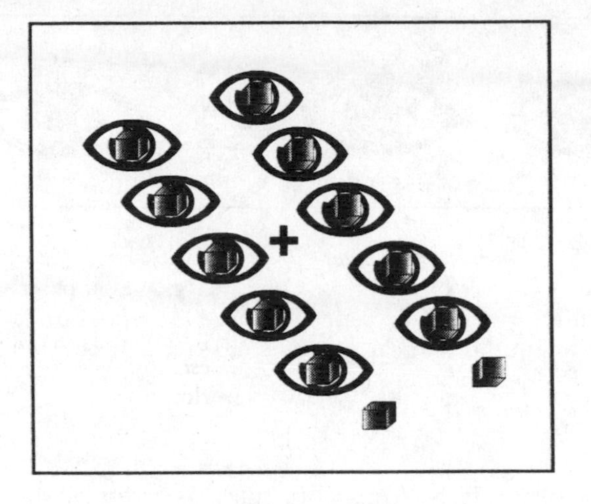

Figure A.4. Many Agreeing Observers Caught in Parallel Worlds.

With so much agreement in each world, little or no overlap exists between the worlds, everything appears normal, and everyone, all *N* of you, agrees.

However, the two universes could still interfere with each other if we could create a proper overlap. Suppose that you are the black-eyed observer's *memory* of the original box. You exist in your brain as an environment capable of registering the direction of the box's perspective. Now suppose you are the gray-eyed observer's memory and you observe the box and the content of your first memory interacting with each other; but you don't observe the box by itself. How would the classical interpretation, Bohr interpretation, and many-worlds interpretation describe this?

In a classical Newtonian world the two memories would agree. The box is either perspective-up, with you both seeing it that way, or perspective-down, with you both seeing it this way. The world is one way or the other. The box is in only one of those states. But before either of you look, there is no way to determine which way the box will appear.

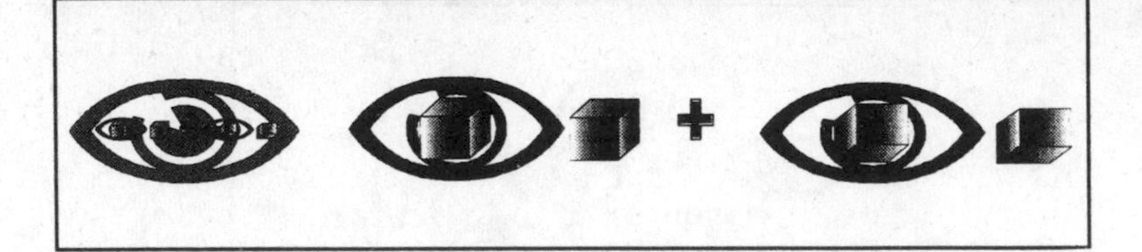

Figure A.5. Gray-Eyed Observer Observes Black-Eyed Observer, while Black-Eyed Observer Observes the Cube. The gray-eyed observer and the black-eyed observer come in contact with the mysterious paradoxical gray pieces/cube. But this time the gray-eyed observer only observes the superposition of both worlds.

According to the quantum rules, as given by the Bohr interpretation, before either of you looks, the box exists in a superposition. After either of you looks, the box will have only one of the two possibilities remaining standing. Regardless of who looks first, both memories will be in agreement.

According to the many-worlds interpretation, in one universe the box is perspective-up and that fact is recorded in your memory 1; in the other universe the box is perspective-down and that fact is recorded in the same memory. And both of these worlds and memories exist in your head. Not only do both exist separately in parallel worlds, but their superposition is also observable by memory 2 in your head. I denote this second observer in you as the pair-of-eyes observer.

The pair-of-eyes observation will be like observing gray pieces, though not quite the same thing since it involves both the box and black-eyed observer's memory. We can call this a pair-of-eyes observation to distinguish it from the gray observation of a single box. Memory 2 records the pair-of-eyes observation.

Suppose now that the gray-eyed observer tells the black-eyed observer about the pair-of-eyes observation and this information enters the black-eyed observer's memory. Then your black-eyed observer memory will contain two very interesting bits of information. In one universe, it will contain the fact that the box is perspective-down and the fact that the box and its parallel universe partner are part of a pair-of-eyes system. In the other universe, it will contain the fact that the box is perspective-up and the fact that the box and its partner are also part of a pair-of-eyes system. Being intelligent, and identifying with one of the perspective views, you will know that a pair-of-eyes observation means that you must have seen the opposite perspective view in a parallel world. You won't be conscious of the opposing box's position, but you will know it exists. Your memory state will be, so to speak, schizophrenic. That pair-of-eyes bit tells you that another parallel universe exists. Your memory, in effect, in this universe, has a "photograph" of another parallel universe. You could think of this other universe as your imaginal realm.

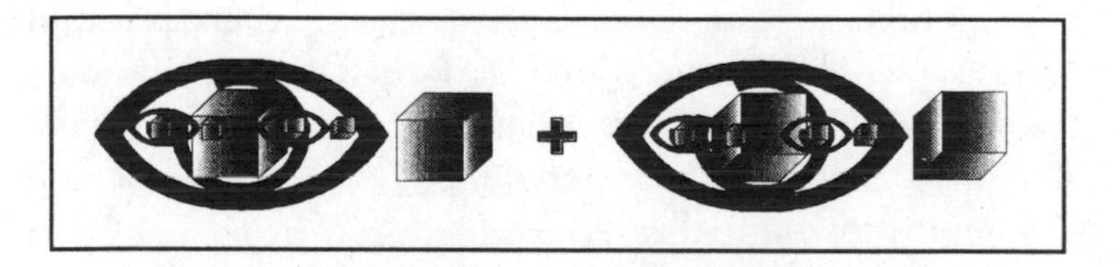

Figure A.6. Black-Eyed Observer Becomes Schizophrenic.

To speculate a bit with this: If we think of your memory 1 as an engram capable of storing a memory, we can construct a model for a well-known psychic disorder called multiple personalities. Each parallel universe engram acts independently unless there is some feedback of information about the totality of personalities. With the feedback about the pair-of-eyes, each engram is aware of the presence of the other parallel engram. This also could be a base explanation for some forms of schizophrenia or altered states of consciousness.

Again, put yourself into the picture as a memory 3 outside of memories 1 and 2. You can see the black-eyed observer and the gray-eyed observer. You can see that the black-eyed observer sees the box and in each world that observer knows what state the box is in. You can see that he also knows that the pair-of-eyes state exists, so you know that he is aware of the other world.

You can also see the gray-eyed observer. She is only aware of the combination of the two worlds. She is only aware of the pair-of-eyes state. If you asked her which way the box was facing, she wouldn't have the slightest idea. In fact, from your point of view, seeing both worlds in parallel, you also can't say which way the box is facing. Of course, if you look at the box and it leaps into one of its states, you will be caught in one of the perspective-up or perspective-down parallel worlds.

Thus, the world depends on you and what you communicate to others. It also depends on what you believe is real. If the many-worlds interpretation is real, then you exist in more than one world and every event in every universe affects you. More than that, you affect everything else in all of these universes in truly countless ways.

Endnotes

Introduction
Awakening the Mystery

1. Today the words Qabalism and Hermeticism are considered to be synonymous terms covering all the arcana and esotericism of antiquity.

2. Manley P. Hall, *The Secret Teachings of All Ages: An Encyclopedic Outline of Masonic, Hermetic, Qabbalistic and Rosicrucian Symbolic Philosophy. Being an Interpretation of the Secret Teachings Concealed within the Rituals, Allegorics, and Mysteries of All Ages* (Los Angeles: The Philosophical Research Society, 1988), p. 114.

3. Eliphas Levi, *Transcendental Magic* (Chicago, 1910), quoted in Stanislas Klossowski de Rola, *Alchemy: The Secret Art* (New York: Thames and Hudson, 1973, 1997).

4. Chris Monnastre, introduction to the 5th edition of *The Golden Dawn*, by Israel Regardie (St. Paul, MN: Llewellyn, 1989), pp. xxii–xxiii.

5. Gershom Scholem, *Major Trends in Jewish Mysticism* (New York: Schocken Books, 1974), p. 22.

6. Ibid., 265, 279.

7. Aleph (א) has, as do the other of the first nine letters, three projections. The first projection is the letter itself. Aleph's other two projections are yod (י) and qof (ק). These projections are forms of the seed-letter aleph. Hence yod is aleph in existence, the symbol for the existent real world of objects we take as "out there." Qof is aleph in its highest realizable state, the so-called exalted or cosmic aleph. Similarly the other letters also have existent and cosmic projections.

8. Dallet (ד) is fundamental resistance and is extremely important, for without it no universe is possible. Dallet's projections are mem (מ) and tav (ת). Mem represents water or the first existent resistance necessary for consciousness to manifest, while tav represents cosmic resistance and is required in order that a single word come into existence.

Chapter 1
Void: The Impossible Life/Death Principle

1. I will be presenting many of these discoveries to you in the course of the book. The basis for all of them is quantum physics.

2. Cherry Gilchrist, *The Elements of Alchemy* (Rockport, MA: Element, 1991), p. 7.

3. I shall be using these terms throughout the text. *Real* refers to that which our common senses tell us and, most importantly, allow us to communicate what we sense to another sentient being. Until that other being tells us that what we sense is also sensed by him or her, what we sense is not real. *Imaginal* is exactly the same as real except there is no other being with whom we share what we sense. Hence anything we sense is primarily imaginal and only becomes real when another confirms it. Anything we sense, we define as *prima materia.*

4. Newton's *law of inertia* essentially states that things tend to stay as they are, resisting any change. Thus a body in motion tends to keep moving as it did earlier. If it was at rest, it tended to stay at rest. If it moved, it tended to move in the same way it moved before along a straight line with constant speed.

5. According to author Barbara G. Walker, *alchemy* in Arabic means "matter of Egypt." Indeed the Arabic *Al-Khemeia* derives from *Khemennu,* "land of the Moon," an old name for Egypt. Long associated with madness or lunacy, the moon has always been associated with the negative aspects within the field of psychotherapy as psychologist Carl Jung noted. Barbara G. Walker, *The Woman's Encyclopedia of Myths and Secrets* (New York: Harper and Row, 1983).

Chapter 2
The Word: Something from Nothing

1. Marshall McLuhan, *The Medium is the Massage: An Inventory of Effects* (San Francisco: Hardwired, 1996).

2. But, just what nothing *is*, is unknown. The best anyone can come up with about nothing is to say it is the negation of anything that exists, anything that is real. Thus it is even the negation of any idea that arises, even the idea of negation.

3. Just as an aside, the Hebrew word for *word* is *aut* (אות—read right to left, aleph-waw-tav). Since each letter is also a word the Hebrew word for *word* is really a code. Spelling out the code reveals aleph, the spirit, acting through vav (sometimes pronounced waw), a fertilizing agent, which disseminates spirit to produce tav, cosmic resistance. Thus a word is God's way of ensuring that resistance will occur. Aleph is the beginning letter of the *alephbayt*, vav means *and*, and tav is the final letter; so aleph-vav-tav also means the beginning and the end—as in a quote, often attributed to John in the Greek version of the Bible, *I am the alpha and the omega.*

4. Evelyn Fox Keller, *Refiguring Life: Metaphors of Twentieth-Century Biology* (New York: Columbia University Press, 1995), p. x.

5. This is extremely important to consider regarding what physicists name fundamental particles. Such names as quarks, coming from the author James Joyce, come to mind. Quarks can have properties including charm and strangeness—merely quantum units—suggesting some property that is constant whenever these particles interact with each other.

6. All possible *classical binary* computers can be envisioned as operating this way and are called *Turing machines* after the English mathematician Alan Turing, who discovered unique ways of cracking codes during World War II. A new field of *quantum computers* (QCs) recently appeared. The QC would not operate in this manner. Instead the QC would somehow both punch a hole and not punch a hole on tape 2 at the same time. A square with such a hole is called a quantum bit or *qubit.*

7. John Archibald Wheeler, "Information, Physics, Quantum: The Search for Links," in *Complexity, Entropy, and the Physics of Information, Santa Fe Institute Studies in the Sciences of Complexity,* vol 8, ed. W. H. Zurich (Redwood City, CA: Addison-Wesley, 1990), p. 5.

8. Modern computers do not use paper tapes (much older models used to). They have "read-write heads" that read or write on the "hard drive," a set of nested magnetically sensitive disks. Disk

technology is quite sophisticated, but nevertheless no different from running a paper tape through a slot and reading a *0* or *1* indicated by a punched hole or its absence.

Chapter 3
The Mind in Body: The Desire to Move

1. Physicist David Bohm first pointed out that complementarity played a role when you began to think about your own thoughts. See David Bohm, *Quantum Theory* (New Jersey: Prentice-Hall, 1951), pp. 169–171.

2. C. G. Jung, *Analytical Psychology: Its Theory & Practice* (New York: Vintage, 1970), pp. 11–15.

3. Paramahansa Yogananda, *Autobiography of a Yogi* (Los Angeles: Self-Realization Fellowship, 1973).

4. Jane Roberts, *The Unknown Reality*, vol. 1 (Englewood Cliffs, NJ: Prentice-Hall, 1977).

5. Fred Alan Wolf, *Star Wave: Mind, Consciousness, and Quantum Physics* (New York: Macmillan, 1984).

6. C. G. Jung, *Analytical Psychology: Its Theory & Practice*, p. 13.

7. Edward F. Edinger, *Anatomy of the Psyche: Alchemical Symbolism in Psychotherapy* (Chicago and LaSalle, IL: Open Court, 1985).

Chapter 4
Inertia: The Mysterious Resistance

1. You probably believe that brains contain minds. Most scientists believe this too. Yet neurologists cannot find the mind in the brain. Sure the mind appears to vanish when someone becomes comatose, just as your car ceases to move when someone pulls all of the spark-plug electrical wiring. We know that the wiring does not drive the car, yet we believe that the neural "wiring" drives the human being. Perhaps the mind vacates comatose brains simply because they don't work, much as a driver leaves a roadside vehicle when it ceases to function.

2. Both *tai chi* and *ch'i* are well-known concepts in China. Briefly, the first deals with a kind of psychic balancing act between body and mind, while the second deals with the means by which an action or transformation can occur.

3. C. G. Jung, *Analytical Psychology: Its Theory and Practice*, pp. 11–15.

Chapter 5
Life: The Body in Mind

1. If I were my body, my subjective *I* would not be separated from my sensations. But I, as you, can clearly distinguish that a sensation arising somewhere in the body is distinct from the I who perceives it. Hence, in some manner I am not my body. In fact from this viewpoint, the arising of a sensation in my body is no different from the arising of sensation outside of my body, as the sensation of the color red in a flower or a musical note, C-sharp, in a symphony. One could argue that these sensations do not arise "out there" but "in here" at my eyes or ears. But I, the perceiver of these sensations, am not my eyes or ears as any blind or deaf person could tell you. Some might argue that I am in my brain and nervous system. However, neurophysiological investigation, while indicating complex neural processing, shows no physical presence of anything like an I. Hence, somehow I am distinct from my sensations and my various organs used to perceive them wherever they may be in the world and wherever they may be believed, imagined, or thought to arise.

2. The Human Genome project maps molecular paths with much success. See the endnotes in chapter 6.

3. We call these paths "histories." It is as if each simpler possible history must somehow contribute to a whole complex actual history. As long as no attempt is made to alter or determine which is the "real" history, all possible histories must be accounted for.

4. The fact that observation takes its toll on matter is one of the main quantum physics ideas of this book. There are several science papers dealing with the observer effect. Please look at the references to Aharonov, Bohm, Goswami, Heisenberg, Herbert, Hoyle, Pauli, Schrödinger, Wheeler, and Wolf in the bibliography.

5. See my articles "The Quantum Physics of Consciousness: Towards a New Psychology," *Integrative Psychology 3* (1985): 236–247; and "On the Quantum Physical Theory of Subjective Antedating," *Journal of Theoretical Biology* 136 (1989): 13–19. Also see my articles "The Dreaming Universe," *Gnosis Magazine* (Winter 1991–92) and "The Timing of Conscious Experience," *Journal of Scientific Exploration* 12, no. 4 (Winter 1998): 511–542.

6. See my book *Taking the Quantum Leap: The New Physics for Nonscientists*, revised edition (San Francisco: Harper and Row, 1981, 1989).

7. Werner Heisenberg, *Physics and Philosophy* (New York: Harper and Row, 1958), p. 41.

8. For those interested in further discussion of parallel worlds ideas as they apply to consciousness, please refer to the appendix and references 2, 5, 6, and 9 in the bibliography.

9. Arnold Mindell, *Dreambody: The Body's Role in Revealing the Self* (Santa Monica, CA: Sigo Press, 1982).

10. George von Békésy, *Sensory Inhibition* (Princeton, NJ: Princeton University Press, 1967).

11. B. Libet, E. W. Wright, B. Feinstein, and Dennis Pearl, "Subjective Referral of the Timing for a Conscious Sensory Experience: A Functional Role for the Somatosensory Specific Projection System in Man," *Brain* 102, part 1 (March, 1979). Also see my own work on this, "The Timing of Conscious Experience," *Journal of Scientific Exploration* 12, no. 4 (Winter 1998): 511–542.

12. See for example, Jim Pouley, *The Secret of Dreaming* (Templestowe, Australia: Red Hen Enterprises, 1988). Also see, Peter Sutton, ed., *Dreamings: The Art of Aboriginal Australia* (Victoria, Australia: Penguin, 1988) and Jean A. Ellis, *From the Dreamtime: Australian Aboriginal Legends* (Australia: Collins Dove, 1991).

13. See Henri Corbin, *Mundis Imaginalis or the Imaginal and the Imaginary* (Ipswich, England: Golgonooza Press, 1976; originally published in spring, 1972).

14. See my book, *The Eagle's Quest: A Physicist's Search for Truth in the Heart of the Shamanic World* (New York: Summit Books, 1991).

Chapter 6
Endless Fertility: Is the Force with Us?

1. Fred Hoyle, *The Intelligent Universe* (New York: Holt, Rinehart, and Winston, 1983), pp. 6–8.

2. Richard Dawkins, *The Blind Watchmaker* (London: Longmans, 1986), p. 316.

3. I can guess that the explanation went something like this: the new electrical atmosphere provided the conditions that allowed organic molecules to form. These molecules in turn were better adapted to survive in such an atmosphere than were simpler molecules.

4. Here we deal with statistics and large number calculations. I refer the reader to chapter 1 of *The Intelligent Universe*, by Fred Hoyle (New York: Holt, Rinehart, and Winston, 1983), pp. 11–23.

5. I discussed this and other experiments in my earlier books. See for example, *The Spiritual Universe: One Physicist's Vision of Spirit, Soul, Matter and Self* (Portsmouth, NH: Moment Point Press, 1999; originally published by Simon and Schuster, 1996), endnotes 18 and 19, pp. 292, 293.

6. Robert Waterson of Washington University and John Sulston of the Sanger Centre in Cambridge, England announced the successful mapping of the complete gene sequences for a whole animal—an entire 97 million genetic "letters" that spell out 19,900 genes—that determines how to make a tiny soil-dwelling worm. With this knowledge, the scientists are able to determine every step of the worm's development from an embryo to a full 959-cell adult containing exactly 302 neurons. This finding gives insight into the long-term process of evolution, suggesting that the genes were present and their functions were already established in the common ancestor of fungi and animals. See the Genome Sequencing Consortium, "Genome sequence of the nematode *C. elegans*: A platform for investigating biology," *Science* 282 (1998): 2,012–2,021.

7. The recently discovered mole rat from Africa appears to be such a DAIFE creature. Living in the ground like a mole with the environment ever presenting a constant temperature, the animal

possesses no ability to shiver when cold or sweat when hot—typical responses to changing thermal environment. It has no need for these adaptations since for this creature it is never cold or hot.

CHAPTER 7
My Time is Your Time: Anything is Possible

1. This interpretation was first given by Niels Bohr when dealing with atomic objects. In the Bohr school, such objects can no longer be observed in the same way as we see the usual world described by Newtonian physics. Instead, objects possess two kinds of complementary observables: those that can be observed simultaneously and those that cannot. Simultaneous observables are called *commuting*; the others are called *complementary*. For example, objects cannot be observed with both positions and momenta simultaneously. Position and momentum are complementary observables. Energy and momentum can sometimes be observed simultaneously, so they are commuting observables. Accordingly, whenever an observable is observed, its complementary observable becomes undefined, while any of its commuting observables are capable of being defined.

2. A real number is any number existing in the range of numbers from $-\infty$ to $+\infty$ (minus to plus infinity). Thus, for example, 5.26 is a real number. So is $-375{,}002.348$. An imaginary number is any real number multiplied by the square root of -1. Since a negative number cannot have a real square root, such numbers as $\sqrt{-1}$ are designated with the symbol i and are called imaginary numbers. Thus $i5$ is the imaginary number 5 and means 5 multiplied by $\sqrt{-1}$. Quantum wave functions are represented by combinations of real and imaginary numbers and cannot be represented by real numbers alone. This fact suffices to prove them imaginal things rather than real things.

3. For those interested in mathematics, ψ^* is mathematically the *complex conjugate* of ψ. Suppose a and b are two ordinary real numbers like 3 and 4. Complex conjugate means that if ψ was the number $a+ib$ (that is, $3+i4$) then ψ^* would be $a-ib$ (that is, $3-i4$). Since $i \times i = -1$, multiplication of these two numbers results in a^2+b^2 ($3^2+4^2=25$), a number that is both real and positive. This number signifies the probability of the event represented by the wave function.

4. First of all I want to recommend two texts that cover basic ideas that arose as slogans from quantum physics: Nick Herbert's excellent *Quantum Reality* (New York: Doubleday, 1985) and my own *Taking the Quantum Leap: The New Physics for Nonscientists*, rev.ed (New York: HarperCollins, 1989). Then I would like to suggest you read my other books: *Star Wave: Mind, Consciousness, and Quantum Physics* (New York: Macmillan, 1984); *Parallel Universes: The Search for Other Worlds* (New York: Simon and Schuster, 1989); *The Dreaming Universe: A Mind-Expanding Journey into the Realm Where Psyche and Physics Meet* (New York: Simon and Schuster, 1994; Touchstone, 1995); and *The Spiritual Universe: One Physicist's Vision of Spirit, Soul, Matter and Self*, rev.ed. (Portsmouth, NH: Moment Point Press, 1999). For more information dealing with physics and consciousness, see Amit Goswami's provocative *The Self-Aware Universe: How Consciousness Creates the Material World* (New York: Tarcher/Putnam, 1993). And for further resources dealing with quantum physics and consciousness, you can go to my web page: www.stardrive.org/fred.shtml.

5. The value 1 means it certainly would occur.

6. Of course deviations from the expected probabilities could be observed. In one hundred experiments the results could be forty-five for 1 and fifty-five for 2 and no one would be concerned. But as the number of experiments grew, say to 10,000 experiments, people would be surprised if the results were 4,500 for 1 and 5,500 for 2, but not surprised if the results were roughly 4,900 for 1 and 5,100 for 2. As the number of experiments increases the deviation from the expected result decreases.

7. Think of a magnetic field. Although it is invisible it nevertheless exerts a force on objects. Also you can consider a gravitational field. You clearly feel its effects on your body as you stumble out of bed in the morning.

8. We imagine that the world of "out there" stuff consists of these tiny whirling-through-space bits of matter. Although we would never attempt to count these bits, we can take it that they behave themselves statistically, always staying within the rules of logic and probability. Thus, the air molecules in the room don't just freeze up in one corner of the room and heat up in another. Although people do report such things from time to time, it is more likely they were only feeling a local draft of cold air sweeping into the room.

9. I discussed these in my article "The Timing of Conscious Experience," *Journal of Scientific Exploration* 12, no. 4 (Winter 1998): 511–542.

10. Many in the field of physics think that conscious experience (CE) indicates "collapse of the wave function" or the dispersion of quantum states or something akin. In this process, something mysterious happens and out of the different possibilities somehow one choice is made.

11. Again, see my article "The Timing of Conscious Experience," *Journal of Scientific Exploration* 12, no. 4 (Winter 1998): 511–542.

12. The *binding problem* refers to the fact that we see a complete and whole picture of the world, not a flash of fragments. Of course one might not expect to see a flash of fragments, but since the brain processes incoming sensory data in different ways and uses different periods of time to do so, one must wonder why the world appears to us so consistent and unfragmented.

13. I suggest that this is the way consciousness works. See my article "The Timing of Conscious Experience." The idea for this article arose from reading several papers including Yakir Aharonov, Peter G. Bergmann, and Joel L. Lebowitz, "Time Symmetry in the Quantum Process of Measurement," *Physical Review* 134B (1964): pp. 1,410–1,416. Also see Lev Vaidman, "Time-symmetrized counterfactuals in quantum theory" (July 1998), available on the internet at http://xxx.lanl.gov/list/quant-ph/new (the Los Alamos National Laboratory reference number is quant-ph/9807075 27).

The time-symmetry model helps us understand all of the paradoxes associated with attempts to map consciousness into spacetime, such as Libet's timing paradox, and those associated with the collapse of the wave function's violations of Lorentz transformational invariance (when collapses are pictured as occurring over spacelike surfaces).

14. "Classical" means here a set of events that can be put together making a plausible story. "What goes up must come down" makes a plausible classical story. "The coin fell heads or it fell tails" sounds reasonable and makes a classical story. Ghost stories are not

classical, although like the story of Scrooge, they may indeed be classical literature. Yet, even here, we, the readers of Dicken's tale, see the plausibility of Scrooge's ghosts.

15. My illustration misses something important. Just as overlapping-cube views make up a single, complementary flat-pieces view, there should also be overlapping-pieces views making one or the other of the cube views. I didn't show that. I didn't know how. Perhaps one of you out there can see how to do this.

Chapter 8
Meaning and Manifestation: The Great Gathering

1. Remember this is the theory from physics (discussed in chapter 5) which posits that besides our own universe there exist alternate parallel universes that come into being whenever anything in our universe interacts with anything else, particularly an observer.

2. In the appendix, I engage you to experimentally enter a parallel-worlds story.

3. Remember there are six common senses: hearing, sight, smell, touch, taste, and mind. The mind sense detects events (called memories) in time, while the other senses detect them in space.

4. Abnormal literally means off the beaten path, far from the norm or average.

5. Light is truly amazing when you consider space, time, and matter from its point of view. According to Einstein's special theory of relativity, no time passes for a light beam as it makes its way across the universe and no space is ever traversed. This is due to the special way in which spatial and temporal intervals transform when one measures them from relatively moving frames of reference.

Chapter 9
Structure and Beauty: Spirit and Soul

1. Such a chunk can be an electron, proton, quark, and so on.

Appendix
Existence: Mind in a Box

1. The word "possibility" is used in two senses here. The first is the more familiar one. In the second sense, it means that a quantum physically defined mathematical quantity when multiplied by itself results in the probability that an event happens. Basically, a possibility squared equals a probability.

2. Everett's ideas are explained in Bryce S. Dewitt, "Quantum Mechanics and Reality," *Physics Today* (September 1970): 30–35.

3. See David Z. Albert, "How to Take a Photograph of Another Everett World," in *New Techniques and Ideas in Quantum Measurement Theory*, ed. D. M. Greenberger, vol. 480, *Annals of the New York Academy of Sciences* (December 30, 1986). Also see "On Quantum-Mechanical Automata," *Physics Letters* 98A, nos. 5, 6 (October 24, 1983): 249–252.

Bibliography

Aharonov, Yakir, Peter G. Bergmann, and Joel L. Lebowitz. "Time Symmetry in the Quantum Process of Measurement." *Physical Review* 134B (1964): 1,410–16.

Bohm, David. *Quantum Theory.* 1951. Reprint, New York: Dover Publications, 1989.

Corbin, Henri. *Mundis Imaginalis or the Imaginal and the Imaginary.* 1972. Reprint, Ipswich, England: Golgonooza Press, 1976.

Crick, Francis. *The Astonishing Hypothesis: The Scientific Search for the Soul.* New York; Charles Scribner and Sons, 1994.

Dawkins, Richard. *The Blind Watchmaker.* London: Longmans, 1986.

Dennett, Daniel G. *Darwin's Dangerous Idea: Evolution and the Meanings of Life.* New York: Touchstone, 1996.

Edinger, Edward F. *Anatomy of the Psyche: Alchemical Symbolism in Psychotherapy.* Chicago: Open Court, 1985.

Einstein, Albert. *Ideas and Opinions.* New York: Crown, 1954.

Einstein, Albert, R. C. Tolman, and B. Podolsky. "Knowledge of Past and Future in Quantum Mechanics." *Physical Review* 37 (1931): 780–81.

Eldridge, Niles. *Reinventing Darwin: The Debate at the High Table of Evolutionary Theory.* New York: John Wiley and Sons, 1995.

Ellis, Jean A. *From the Dreamtime: Australian Aboriginal Legends.* Australia: Collins Dove, 1991.

Gilchrist, Cherry. *The Elements of Alchemy.* Rockport, MA: Element, 1991.

Goswami, Amit. *The Self-Aware Universe: How Consciousness Creates the Material World.* New York: Putnam Books, 1993.

Heisenberg, Werner. *Physics and Philosophy.* New York: Harper and Row, 1958.

Herbert, Nick. *Quantum Reality.* New York: Doubleday, 1985.

Hesse, Hermann. *Siddhartha.* Translated by Hilda Rossner. New York: New Directions, 1951.

Holmyard, E. J. *Alchemy.* Mineola, New York: Dover, 1990.
Hoyle, Fred. *The Intelligent Universe.* New York: Holt, Rinehart, and Winston, 1983.
Jung, C. G. *Analytical Psychology: Its Theory and Practice.* New York: Vintage, 1970.
Kauffman, Stuart. *At Home in the Universe: The Search for Laws of Self-Organization and Complexity.* New York: Oxford University Press, 1995.
Keller, Evelyn Fox. *Refiguring Life: Metaphors of Twentieth-Century Biology.* New York: Columbia University Press, 1995.
Klossowski de Rola, Stanislas. *Alchemy: The Secret Art.* New York: Thames and Hudson, 1973, 1997.
Libet, B., E. W. Wright, B. Feinstein, and Dennis Pearl. "Subjective Referral of the Timing for a Conscious Sensory Experience: A Functional Role for the Somatosensory Specific Projection System in Man." *Brain* 102, part 1 (March 1979).
Lwoff, André. *Biological Order.* Cambridge, MA: MIT Press, 1962.
McLuhan, Marshall. *The Medium is the Massage: An Inventory of Effects.* San Francisco: Hardwired, 1996.
Mindell, Arnold. *Dreambody: The Body's Role in Revealing the Self.* Santa Monica, CA: Sigo Press, 1982.
Pauli, Wolfgang. "Ideas of the Unconscious from the Standpoint of Natural Science and Epistemology." *Dialectica* 8, no. 4 (December 1954).
———. "Science and Western Thought." In *Europa: Erbe und Auftrag,* edited by M. Gohring. Mainz, Germany: Internationaler Gelehrtenkongress, 1955.
———. Letter to C. G. Jung (March 1953). Exhibited in the Pauli Room at CERN, Geneva.
Pouley, Jim. *The Secret of Dreaming.* Templestowe, Australia: Red Hen, 1988.
Price, Huw. *Time's Arrow and Archimedes' Point.* New York: Oxford University Press, 1966.
Roberts, Jane. *The Unknown Reality.* Vol 1. 1977. Reprint, San Rafael, CA: Amber-Allen, 1996.
Rosen, Eliot Jane, ed. *Experiencing the Soul: Before Birth, During Life, After Death.* Carlsbad, CA: Hay House, 1998.
Schiff, Leonard I. *Quantum Mechanics.* 3rd ed. New York: McGraw-Hill, 1955, 1968.

Scholem, Gershom. *Major Trends in Jewish Mysticism.* New York: Schocken Books, 1974.

Schrödinger, Erwin. *My View of the World.* Cambridge, England: Cambridge University Press, 1964. Reprint, Woodbridge, CT: Ox Bow Press, 1983. Originally published in German (Hamburg-Vienna: Paul Zsolnay Verlag, 1961).

———. *What is Life? & Mind and Matter.* Cambridge, England: Cambridge University Press, 1967.

Smoley, Richard. "My Mind Plays Tricks on Me." *Gnosis* (Spring, 1991): 12.

Suarés, Carlo. *The Cipher of Genesis: The Original Code of the Qabala as Applied to the Scriptures.* Berkeley, CA: Shambhala, 1970. Reprint, York, ME: Samuel Weiser, 1992.

———. *Les Spectrogrammes de L'Alphabet Hebraïque.* Geneva, Switzerland: Mont-Blanc, 1973.

———. "The Cipher of Genesis." In *Tree 2: Yetzirah,* edited by David Meltzer. Santa Barbara, CA: Christopher Books, 1971. From a lecture by Suarés reprinted from *Systematics* 8, no. 2 (September 1970).

———. *The Second Coming of Reb Yhshwh.* York Beach, ME: Samuel Weiser, 1994.

Sutton, Peter, ed. *Dreamings: The Art of Aboriginal Australia.* Victoria, Australia: Penguin, 1988.

Vaidman, Lev. "Time-Symmetrized Counterfactuals in Quantum Theory." On the internet at http://xx.lanl.gov/ (quant-ph/9807075 July 27, 1998).

von Békésy, George. *Sensory Inhibition.* Princeton, NJ: Princeton University Press, 1967.

von Franz, Marie-Louise. *Number and Time: Reflections Leading to a Unification of Depth Psychology and Physics.* Evanston, IL: Northwestern University Press, 1974.

———. *Alchemical Active Imagination.* Boston: Shambhala, 1997.

Walker, Barbara G. *The Woman's Encyclopedia of Myths and Secrets.* New York: Harper and Row, 1983.

Waterson, Robert and John Sulston. "Genome Sequence of the Nematode *C. Elegans:* A Platform for Investigating Biology." *Science* 282 (1998); 2,012–021.

Watzlawick, Paul. *How Real is Real?* New York: Random House, 1976.

Weiner, Norbert. *God and Golem, Inc: A Comment on Certain Points Where Cybernetics Impinges on Religion.* Cambridge, MA: MIT Press, 1964.

Wheeler, John Archibald. "Information, Physics, Quantum: The Search for Links." In *Complexity, Entropy, and the Physics of Information, Santa Fe Institute Studies in the Science of Complexity* 8, edited by W. H. Zurich. Redwood City, CA: Addison-Wesley, 1990.

———. "How Come the Quantum?" In *New Techniques and Ideas in Quantum Measurement Theory,* edited by D. M. Greenberger. Annals of the New York Academy of Sciences 480 (December 30, 1986).

Wolf, Fred Alan. *Taking the Quantum Leap: The New Physics for Nonscientists.* San Francisco: Harper and Row, 1981. Revised, New York: HarperCollins, 1989.

———. *Star Wave: Mind, Consciousness, and Quantum Physics.* New York: Macmillan, 1984.

———. "The Quantum Physics of Consciousness: Towards a New Psychology," *Integrative Psychology* 3 (1985): 236-47.

———. *The Body Quantum: The New Physics of Body, Mind, and Health.* New York: Macmillan, 1986.

———. "The Physics of Dream Consciousness: Is the Lucid Dream a Parallel Universe?" *Lucidity Letter* 6, no. 2 (December 1987): 130–35.

———. *Parallel Universes: The Search for Other Worlds.* New York: Simon and Schuster, 1989.

———. "On the Quantum Physical Theory of Subjective Antedating." *Journal of Theoretical Biology* 136 (1989): 13–19.

———. *The Eagle's Quest: A Physicist's Search for Truth in the Heart of the Shamanic World.* New York: Summit, 1991.

———. "The Dreaming Universe." *Gnosis* 22 (winter 1992): 30–35.

———. *The Dreaming Universe: A Mind-Expanding Journey into the Realm where Psyche and Physics Meet.* New York: Simon and Schuster, 1994. Reprint, New York: Touchstone, 1995.

———. "The Body in Mind." *Psychological Perspectives* 30 (fall-winter 1994): 22–35.

———. "The Quantum Mechanics of Dreams and the Emergence of Self-Awareness." In *Toward a Scientific Basis for Consciousness,* edited by S. R. Hameroff, A. W. Kaszniak, and A. C. Scott. Cambridge, MA: MIT Press, 1996.

———. *The Spiritual Universe: One Physicist's Vision of Spirit, Soul, Matter, and Self.* Portsmouth, NH: Moment Point Press, 1999. Originally published as *The Spiritual Universe: How Quantum Physics Proves the Existence of the Soul.* New York: Simon and Schuster, 1996.

———. "The Soul and Quantum Physics." In *Experiencing the Soul: Before Birth, During Life, After Death,* edited by Eliot Jay Rosen. Carlsbad, CA: Hay House (1998): 245–52.

———. "The Timing of Conscious Experience." *Journal of Scientific Exploration* 12, no. 4 (winter 1998): 511–42.

———. "A Quantum Physics Model of the Timing of Conscious Experience." In *Toward a Science of Consciousness III: The Third Tucson Discussion and Debates,* edited by Stuart Hameroff, Al Kaszniak, and David Chalmers. Cambridge, MA: MIT Press, 1999.

———. "The Quantum Physical Communication Between the Self and the Soul." *Noetic Journal* 2, no. 2 (April 1999): 149–57.

Yogananda, Paramahansa. *Autobiography of a Yogi.* Los Angeles: Self-Realization Fellowship, 1973.

Zohar: The Book of Enlightenment. Translated and introduction by Daniel Chanan Matt. Mahwah, NJ: Paulist Press, 1983.

INDEX

"Absolute Realization," 3
action and reaction, 136
actualities vs. possibilities, 101, 115, 139, 143, 146, 150, 164
Adam Kadmon, 5, 6, 10, 24, 131–33
alchemical transformation, 32
alchemist, 18, 19, 20, 22, 39, 133, 135
alchemy, 2, 3–4, 17, 19
  ancient, 2, 18, 40, 49, 132
  ancient laws of movement in, 39–41
  fifth essence *(quintessence)* of, 41
  and spiritual transformation, 49
aleph, 9, 13, 25, 37, 51, 65, 82, 131, 141, 153
amino acids, 94
anatomy of the psyche, 49
ancient (old) alchemists, 2, 6, 19, 20, 22–24, 31, 40–41, 50
  and qwiffs, 105
antimatter, 47
archetype, 21, 135
Aristotle, 39
atoms, 68–69, 72
aura, 122–23, 138
Austin, J. L., 31
axiom 1 (thinking about thinking), 42–45
axiom 2 (thinking about sensing), 45–46
axiom 3 (thinking about feeling), 46–47
axiom 4 (thinking about intuiting), 47–50
axioms as tools, 58–59

Babylonians, 3
bayt, 9, 25, 37, 51, 65, 82, 129
Bible, 29
Big Bang, 29, 133
biology, 1, 89, 92–94
bit, 31–35
  and bootstrap problem, 35–36
  and computers, 32–34
  definition of, 32–33
  and "in here," 49
  and inner world, 32
  it from, 34–35, 88
  and "out there," 49
body, 15, 68–83, 77–78
  Arnold Mindell on the, 78
  and "dream body," 78
  and the imaginal realm, 34
Bohr, Niels, 104, 144, 147, 149, 160
bootstrap problem, 35–36
Bruce, Lenny, 30
Buddhism, 54, 78, 80
business survival and Darwinism, 91–92

Chaldeans, 3
Chardin, Peirre Teilhard de, 119
chemistry, 3, 6
Christian, 3, 78
complimentarity principle, 73–74
  See also quantum physics
computer, 1, 32–34
consciousness and unconsciousness, 3, 10, 13, 16, 20, 25, 62, 68, 69, 70, 71, 79–82, 87, 112–114, 128, 132–134, 152, 162
Corbin, Henri, 82–83
creation, 4, 9, 24, 50, 57, 62, 131–132
  and God, 35–36
  and laws of transformation, 61
  and resistance, 10, 60–63
  theories of, 29
cycle, 20–21

DAIFE/AIFE, 98, 159
dallet, 10, 51, 65, 82, 131, 153
Darwinism (natural selection), 89–94, 97–99, 133
  and divine right, 92
  and industrial age, 89–92
  and morality, 89–90
  and populist revolutions, 92
On the Origin of the Species (Darwin), 91
Dawkins, Richard, 93
death, 15, 45
déjà vu, 24
Democratus, 69
desire
  four axioms of, 42–50
  and intuition, 48
  and time loops, 106, 116
determinism, 15, 56, 73
divine existence, 5
DNA, 68–69, 93–94, 97–98
dogma, 1, 3
Dreambody (Mindell), 78
dreams, 8, 23, 28, 78, 104
dreamtime, 82–83
duality, 15, 139

earth, 15, 16
Edinger, Edward F., 49
  and anatomy of the psyche, 49
Egyptians, 3
Einstein, 119, 143, 163
  theory of relativity, 22, 104
electrons, 43–47, 68–69, 107–11, 114
elements, 15, 16, 39–41
Empedocles, 39
energy, 13, 29, 46, 48, 68–70, 85, 117, 126, 129, 137
engram, 151
enzymes, 94
Escher, Maurits, 56–57
Everett, Hugh, 119, 145
evil, 18

feeling, 74–75
  and axiom 3, 46–47
  and base feelings, 46
  and electrons, 46–47
  and energy, 46–47
  and matter, 46–47
  and photons, 46–47
fertility and Darwinism, 87–99
free will, 15
Freemasonry, 2
future, 8, 56, 77–78, 95–99, 120–21

gamma rays, 47
ghimel, 10, 37, 51, 65, 82
God, 5, 16, 20, 29, 35, 48, 82, 87, 88, 128, 131, 135, 138, 155
God's great trick, 15–16
goldmaking, 3
"Great Work," 3–4
Greece, 3
group ego, 46

hay, 10, 65, 82
heaven, 15
Hebrew letters, 4, 5, 9–10, 13, 25, 37
  aleph, 9, 13, 25, 37, 51, 82, 131, 141, 153
  bayt, 9, 25, 37, 51, 82, 129
  dallet, 10, 51, 82, 131, 153
  ghimel, 10, 37, 51, 82
  hay, 10, 65, 82
  hhayt, 10, 117, 129
  tayt, 10, 129
  vav, 10, 85
  waw, 10, 85
  zayn, 10, 101, 117
Heisenberg, 71–72
Hermeticism, 2
Hesse, Herman, 103
hhayt, 10, 117, 129
How to Do Things with Words (Austin), 31
Hoyle, Sir Fred, 89

illusion, 16, 17
imaginal/real. *See* real/imaginal
"in here," 16, 19, 20, 24, 28–30, 31–32, 34, 48–50, 74, 79, 96, 107, 112–13, 116, 132
individuation, 46
inertia, 19–21, 53–63, 153
  and mind, 53
  and Newton, 153
  and resistance, 56
  and rest mass, 136
  and thinking, 53–54
information, 16, 21, 22, 28 29–30, 50, 57, 63, 88

from the future, 8, 77–78, 95–99
information theory, 2
inseparability, 18
instincts, 36, 39
intuiting and intuition, 20, 48, 58, 59, 68
axiom 4, 47–50
Carl Jung and, 47, 49
isolation, 54–55
and dukkha, 54

John of Rupescisia, 19
Judaism, 2
Jung, Carl, 5, 20, 45, 47, 49, 58, 74, 154

Kazantzakis, Nikos, 131
Keller, Evelyn Fox, 30

Last Temptation of Christ, The (Kazantzakis), 131
lebensraum and Darwinism, 91
Lenin, Vladimir, 53, 56
Levi, Eliphas, 4
Libet, Benjamin, 79, 112, 160
life/death principle, 15, 17–21
light, 6, 24, 136–39, 163
loneliness, 54
Luke, 15
lunacy, 22–24, 50, 154
Luria, Isaac, 5

magnetic field, 107–10
magic, 23
Magic Mountain, The (Mann), 67, 87
magician, 16, 50, 61–62, 125, 137–38
magnetic field, 109
Magnum Opus, 3, 4
Mann, Thomas, 67, 87
material, 65
matter, 37, 46, 53, 65, 68–70, 88, 104, 134, 135–36, 157
building blocks of, 43–47, 68–71, 93–94, 107–10, 111, 114
and Darwin, 48
force of matter on mind, 54–55
and light, 46
and mind, 8, 53–54
and "spin–up" and "spin–down" theory, 107–09
and Stern–Gerlach device, 109–10
and theory of relativity, 104
and transformation, 48
McLuhan, Marshall, 28
The Medium is the Massage (McLuhan), 28
mechanical laws of motion, 20
memory, 75, 77, 79, 80, 82, 99, 112, 114, 137–38, 149–51
metaphysical, 22
Middle Ages, 3
millennium, 1
Miller, Stanley, 93–94
and DNA, 93–94, 97–98
mind, 15, 19, 37–50, 53, 59, 70, 128, 138, 143–52
body in, 67–83
in body, 8, 39–50
and brain, 156
force of mind on matter, 54
into matter, 10, 21, 54
and objectivism, 96–97
mind–body, 8, 17, 68, 80
mind–objects, 23–24
Mindell, Arnold, 77
molecular evolution, 94
molecular science, 69
molecules, 45, 68–69, 94
Monnastre, Cris, 4
MSE (mapped sequence of experience), 43–44
mystery, 1–11, 16, 108
mysticism, 21, 61
mythology, 5

neurobiology, 2
new alchemists, 7, 41
and spirit, 10, 41, 131–39
and desire, 41–42
new alchemy, 9, 15, 23, 32, 41, 82, 132, 139
and alchemical creativity, 55
and alchemical possibility, 121–22, 127–28
author's first experience with, 6
and complementary ways, 44
definition of, 29–30
and desire, 106
and perception, 32
and physics and psychology, 54, 71, 78
Newton, Isaac, 20, 69, 143, 149, 153, 160

Noh drama, 121–25
NSE (new sequence of experiences), 43–44

objective materialism, 6
objectivism, 96
  and reality, 96
observation and observer, 73, 75–77, 80, 96, 97, 104, 111–16, 119, 121, 143–44, 146–47, 149, 150, 157
old alchemists. *See* ancient alchemists
On the Origin of the Species (Darwin), 91
origin, theories of. *See* creation
"out there," 16, 19, 20, 24, 28–30, 31–32, 34, 48–50, 74, 79, 96, 105, 107, 112–15, 116, 132, 134, 161

Pagels, Heinz, 27,
paradoxical cube, 73–77, 144, 147
parallel universes, 67, 75, 76–77, 81–83, 109–10, 115, 119–21, 136–37, 145, 148–49, 152
particles, 45, 70, 71, 107–08
  possible physical, 71
paths of action, 23
perception, 31–32, 49–50, 79, 96
  See also reality
personal history, concept of, 113
perspective, 146, 148–50
Pert, Candace, 39
philosophy and philosophers, 3, 60, 96, 104
Phoenicians, 3
photons, 46
physical laws of nature, 18
prima materia, 50
principle of complementarity, 72–74, 160
  See also quantum physics
principle of indeterminism, 71–74, 101
  See also quantum physics
probability, 71, 78, 95, 104, 105–07
projections, 153
proteins, 93–94
psi, 105
psi–star, 105
psychology, 54, 71, 78

Qabala, 2
Qabalism, 2–3, 5–6
qof, 10, 132, 153
quantum computers, 155
quantum leap, 10, 67, 75, 125–26
quantum physics, 2, 7–8, 31–32, 44–45, 49, 50, 68–75, 88, 107, 115, 133, 143–44
  and Bohr, Niels, 104, 144, 147, 149, 160
  Copenhagen interpretation, 104, 144
  and Heisenberg, 71–72
  and information from the future, 96–99
  and many worlds interpretation, 146–47, 150
  and parallel universes, 67, 75, 76–77, 81–83, 107–09, 119–21, 136–37, 145, 148–49, 152
  and principle of complementarity, 72–74
  and principle of indeterminism, 71–74
  and time, 44–45, 133
  and uncertainty principle, 71–74
  and wave functions, 105
qwiffs, 105
  and endpoints, 105

real and negative numbers, 160
real/imaginal, 8, 19, 20, 21, 54–55, 57, 60, 62, 63, 76, 81, 82–83, 109, 113, 114–15, 136, 139, 151, 153
reality, 21, 22–23, 32, 50, 57, 137, 139
  actual experiences of, 58–59
  consensus, 148
  "creating your own," 106–08
  as inner world, 32
  and mind–body, 8, 17, 68, 80
  mythic, 70–71
  objective, 8, 23, 32, 80, 113
  and objectivism, 96

and parallel universes, 67, 75, 76–77, 81–83, 107–09, 119–21, 136–37, 145, 148–49, 152
preconceived notions of, 58–59, 61
and quantum physics, 107–08, 152
and qwiffs, 105
and scripts, 77–79, 81–83
subjective, 3, 6, 8, 32, 82, 111–16
and time loops, 104–06, 116
and trickster, 59
virtual, 22–23
Refiguring Life (Keller), 31
Regardie, Israel, 4
reproduction. *See* fertility and Darwinism
resistance, 10, 20–21, 51, 53–63, 87, 153
and the Borg, 55
and causality, 55
and universe, 55
Roberts, Jane, 45
and Seth, 45
Unknown Reality, The, 45
Rosicrucianism, 2

science and scientists, 1, 2, 3, 6, 18, 22, 29, 61, 68, 69, 89–92
script, 77–80, 82
Sensing and sensations, 45–46, 58, 79, 157
axiom 2, 45–46
and feelings, 68
and perception, 49–50
and spatial extent, 45
separability, 18, 19
Seth, 45
Jane Roberts and, 45
Unknown Reality, The and, 45
shadow, 5
Siddhartha (Hesse), 103
Sixpack, Joe, 121–25
soul, 131–39
space and time, 17, 22, 37, 70, 71, 73, 76, 79, 80, 82, 87, 96, 104, 105, 107, 123–25, 128, 136, 162, 163
spatial extent, 45
spatial temporal realm, 22
species survival, 89–90
spirit, 6, 25, 37, 65, 133, 135
spiritual, 25
spirituality, 22
spiritual revelation, 5
Star Wave (Wolf), 46
"states," 71
subatomic particles, 68

TAT (thinking about thinking), 43–44
tayt, 10, 129
The Golden Dawn (Regardie), 4–5
The Medium is the Massage (McLuhan), 28
The Prisoner, 27–28
theory of relativity, 42, 104
thinking, 53, 74–75
axiom 1, 42–43
axiom 2, 45–46
axiom 3, 46–47
axiom 4, 47–50
and MSE, 43–44
and NSE, 43–44
and TAT, 43–44
and theory of relativity, 42
time and space, 17, 22, 37, 70, 71, 73, 76, 80, 82, 87, 96, 104, 105, 123–25, 128, 136, 162
time loop, 104–06
time
and MSE, 43–44
nature of, 70–71, 141
and NSE, 43–44
relativity of, 43
space and, 17, 37, 70, 71, 73, 76, 80, 82, 87, 96, 104, 105, 123–25, 128, 136, 162
stop time, 124
timewind, 88
Transcendental Magic (Levi), 4
trickster, 23, 50, 57, 61–62, 67
and perceptions of reality, 59

uncertainty principle, 73–74
See also quantum physics
unconsciousness. *See* consciousness and unconsciousness

unity, 15
universe, 15, 17, 27, 29, 39, 45, 69, 73, 88, 97, 124, 126, 128, 136, 146, 149, 150, 151–52, 163
  and body and mind, 8–9
  and parallel realities and worlds, 67, 75, 76–77, 107–09, 119–21, 136–37, 145, 148–49,
Unknown Reality, The (Roberts), 45
  and Seth, 45
Urey, Harold, 93–94
  and DNA, 93–94

vav, 10, 85, 101
Void, the, 15–24, 29
von Békésy, George, 79–80
Wheeler, John A., 34, 88, 145
words, 25–36
  vs. action, 31
  as hurtful, 30–31
  and minority rights, 30–31
Word of God, the, 29

yod, 10, 141
yoga, 45
Yogananda, Swami, 45

zayn, 10, 101
Zeno effect, 96–97

About the Author

Fred Alan Wolf, a Ph.D. in theoretical physics, is a consulting physicist and American Book Award-winning writer. He travels throughout the United States and the world presenting lectures on consciousness and the new physics. If you are interested in attending one of these events or would like to inquire as to Dr. Wolf's availability to speak at your event, please contact him by mail or email:

Dr. Fred Alan Wolf
c/o Moment Point Press
P.O. Box 920287
Needham, MA 02492

fred@fredalanwolf.com
www.fredalanwolf.com

Other Books of Interest from Moment Point Press

The Spiritual Universe
one physicist's vision of
spirit, soul, matter, and self
Fred Alan Wolf, Ph.D.

Lessons from the Light
what we can learn from
the near-death experience
Kenneth Ring, Ph.D.

Speaking of Jane Roberts
remembering the author of
the Seth material
Susan M. Watkins

Consciously Creating Each Day
a 365 day perpetual calendar of
spirited thought
from voices past and present
edited by Susan Ray

Three Classics in Consciousness
by Jane Roberts
Adventures in Consciousness
Psychic Politics
The God of Jane

for more information
visit us at
www.momentpoint.com